Ecological Relationships and Evolution of the Rickettsiae

Volume I

Author

Nyven J. Marchette, Ph.D.
Professor of Tropical Medicine and Public Health
The John A. Burns School of Medicine
University of Hawaii
Honolulu, Hawaii

With contributions by

David Stiller, Ph.D.
Hemoparasitic Diseases Research Unit
Hemoparasite Laboratory
Agriculture Research Service
U.S. Department of Agriculture
Washington State University
Pullman, Washington

CRC Press
Taylor & Francis Group
Boca Raton London New York

CRC Press is an imprint of the
Taylor & Francis Group, an **informa** business

First published 1982 by CRC Press
Taylor & Francis Group
6000 Broken Sound Parkway NW, Suite 300
Boca Raton, FL 33487-2742

Reissued 2018 by CRC Press

Library of Congress Cataloging in Publication Data

Marchette, Nyven J., 1928–
Ecological relationships and evolution of
the rickettsiae.

Bibliography: p.
Includes index.
1. Rickettsia. 2. Microbial ecology.
3. Rickettsia—Evolution. I. Stiller, David,
1931– . II. Title.
QR353.M37 576'.62 81-18005
ISBN 0-8493-6125-7 (v. 1) AACR2
ISBN 0-8493-6126-5 (v. 2)

A Library of Congress record exists under LC control number: 81018005

ISBN 13: 978-1-315-89247-4 (hbk)
ISBN 13: 978-1-351-07157-4 (ebk)

Visit the Taylor & Francis Web site at http://www.taylorandfrancis.com and the
CRC Press Web site at http://www.crcpress.com

PREFACE

"What song the Syrens sang, or what name Achilles assumed when he hid himself among women, although puzzling questions, are not beyond all conjecture." In these words Sir Thomas Browne sets no limit to speculation, but it is generally admitted that speculation is idle if it is useless. The present review of rickettsial ecology and evolution assumes that speculation should be carried as far as is necessary to form a working hypothesis, upon the framework of which both the investigator and the student may hang their ideas. Moreover, a detailed examination of the evolution and possible relationships of the pathogenic rickettsiae may provide the fundamental basis for a natural classification of the family Rickettsiaceae and perhaps even the order Rickettsiales.

The ecological relationships of the rickettsiae for the most part are well known or at least amenable to investigation, but many details relating directly to rickettsial evolution are missing or incompletely known and cannot be subjected to rigorous scientific analysis. Fortunately, however, rickettsiae are obligate parasites with complex life cycles dependent upon certain mammals and arthropods about which a great deal is known. Our knowledge of the present geographical distribution of vertebrates, acarines, and pathogenic rickettsiae is especially good, and immunological and biochemical techniques enable us to establish degrees of relationship between microorganisms with some confidence. Many aspects of the relationships of rickettsiae to their hosts are also well known. The paleontological evidence for the evolution of the mammalian hosts of ticks and their rickettsiae, though far from complete, is sufficient to satisfy most critics. The geological record of changes of the Earth over the past 100 million years or so is likewise well documented, if not complete in every detail. Ticks and mites are almost completely absent from the fossil record, but studies of the systematics and distribution of living forms provide valuable, if incomplete, information on acarine phylogeny. Judicious use of the considerable amount of material that is available relative to rickettsiae should enable one to speculate from a firm base as to the probable course of rickettsial evolution.

An axiom of science is that, in the study of natural phenomena, the simplest explanation that logically fits all the data available is likely to be closest to the truth. This assumes, of course, that all or most of the relevant information concerning the phenomenon under investigation is known, but this is not always the case, and seldom is it so in the study of evolution. Thus, it is essential not to construct too rigid a framework for the rickettsial evolutionary tree. The accumulation of new data may require its periodic pruning, with prudent cutting and grafting of branches here and there, to maintain its viability and preserve its natural symmetry.

In the present study, many fields and disciplines are explored for evidence pertinent to an evolutionary history of the rickettsiae. From these data may emerge a logical sequence of events occuring through geological and historical time, culminating in the rickettsial species living today. As you will see, in certain areas so little is known that the conclusions are scarcely more than educated guesses. Perhaps the most valuable contribution of this study will be in pointing out what is not known rather than in reviewing what is known of rickettsial evolution. If even a few students are stimulated to explore some of the many unsolved problems, this work will have served a useful purpose.

Nyven Marchette

THE AUTHOR

Nyven J. Marchette, Ph.D., is Professor of Tropical Medicine and Public Health at the University of Hawaii, John A. Burns School of Medicine, Honolulu, Hawaii.

He received a Bachelor of Science degree in general zoology from the University of California, Berkeley in 1950 and a Master of Science degree in Invertebrate Zoology from the same institution in 1953. The next two years were spent in the U.S. Army as an enlisted man assigned to the Biological Warfare testing facility at Dugway, Utah where he participated in ecological research in collaboration with a group from the University of Utah. Following discharge from the army, he stayed with the university group and received the Ph.D. degree in bacteriology from the University of Utah, Salt Lake City in 1960.

In 1960–1961 Dr. Marchette was appointed Research Bacteriologist at the University of Utah Ecological Research facility at Dugway Proving Ground, Utah. Between 1961 and 1964 he was a post doctoral fellow and then Assistant Research Microbiologist in the International Centers for Medical Research and Training Program at the University of California, San Francisco and the Institute for Medical Research in Kuala Lumpur, Malaysia. From 1965 to 1967, he moved to the Department of Microbiology in the new University of Malaya Medical School and established the University of California ICMRT Arbovirus Research Laboratory. He remained with the University of California until 1970 running the arbovirus research laboratory in San Francisco at the Hooper Foundation (still associated with the ICMRT).

In 1970, Dr. Marchette accepted a position as Associate Professor of Tropical Medicine and Public Health at the Department of Tropical Medicine and the School of Public Health at the University of Hawaii, Honolulu. From 1974 to the present his title has been Professor of Tropical Medicine and Public Health.

He is a member of The American Society for Microbiology and is currently president of the local Hawaii Branch. His other memberships include the American Society of Tropical Medicine and Hygiene, the American Association for the Advancement of Science, and Sigma Xi.

Dr. Marchette has been Investigator or Associate Investigator on numerous NIH and WHO grants studying viral pathogenesis (principally dengue viruses) and currently is Co-investigator on an NIH grant to determine the etiology of Kawasaki Disease.

In 1977–1978 he was awarded a Fogarty International Senior Fellowship at the John Curtin School of Medical Research, Australia National University, Canberra, Australia. Part of that year was spent in research on the material for this book.

Dr. Marchette's current research interests include the etiology and pathogenesis of viral and rickettsial infections, the ecology of *Ehrlichia* and related rickettsiae in Hawaii, and the epidemiology of virus diseases in Hawaii and the Pacific Basin.

Dr. Marchette has published over 50 articles in the field of Microbiology.

Dedicated to J. Ralph Audy, Ecologist, philosopher, mentor, and friend.

ACKNOWLEDGMENTS

My thanks to all those who read and commented on portions of the manuscript while it was in preparation. I am especially grateful to Dr. Harry Hoogstraal and Dr. Robert Traub for invaluable critical comments. The views expressed in this work, however, are entirely my own, and I am wholly responsible for any errors of fact that may occur. Thanks also to Ms. Iris McCrea for verifying many of the references and to Ms. Francis Kramer, who did the bulk of the final typing.

Ecological Relationships and Evolution of the Rickettsiae

Nyven J. Marchette

Volume I

Introduction
The Distribution and Relationships of Rickettsiae
Evolution and Distribution of Rodents
Evolution of the Tick-Rickettsia Relationships
The Tickborne Rickettsiae of the Spotted Fever or Tick-Typhus Group
The Typhus Complex: *Rickettsia typhi* and *R. prowazekii*. Adaptation to Insects
Index

Volume II

Rickettsia tsutsugamushi—An Acarine Offshoot or a Separate Line?
The Wolbachieae
Origin and Evolution of the Ehrlichieae
Coxiella burnetii—Its Origin and Distribution
The Anaplasmataceae, Bartonellaceae, and *Rochalimaea quintana*
The Chlamydiales
A Natural Classification of the Rickettsiae
Index

TABLE OF CONTENTS

Volume I

Chapter 1
Introduction 1
Nyven Marchette

Chapter 2
The Distribution and Relationships of Rickettsiae 11
Nyven Marchette

Chapter 3
Evolution of the Tick-Rickettsia Relationship 29
Nyven Marchette

Chapter 4
Evolution and Distribution of Rodents 53
Nyven Marchette

Chapter 5
The Tickborne Rickettsiae of the Spotted Fever
or Tick-Typhus Group 75
Nyven Marchette

Chapter 6
The Typhus Complex: *Rickettsia typhi* and
R. prowazekii. Adaptation to Insects 113
Nyven Marchette

Index 155

Chapter 1

INTRODUCTION

N. Marchette

TABLE OF CONTENTS

I. Bacterial Classification 2

II. Evolution of Rickettsial Classification 3
- A. Arthropod Association and Pathogenicity 3
- B. Immunological Characteristics 4
  - 1. The Louseborne and Fleaborne (OX 19-Agglutinating) Group 4
  - 2. The Trombiculid Miteborne (OX K-Agglutinating) Group 5
  - 3. The Tickborne "Spotted Fever" Group 5
- C. The *Proteus* Nonagglutinating Organisms 6

III. Phylogenetic Relationships and Classification According to Vector 6
- A. Terminology 6
- B. Tickborne Rickettsiae 7
- C. Insectborne Rickettsiae 8
- D. Trombiculidborne Rickettsiae 8

IV. Natural Classification of the Rickettsiae 8

References 9

I. BACTERIAL CLASSIFICATION

Cowan[1] correctly pointed out that without a fossil record much of bacterial taxonomy is a ". . . mixture of speculation and wishful thinking." However, I do not agree completely with his claims that ". . . the 'best' classifications are those made by artists with the keenest appreciation of what is both useful and intellectually satisfying." In my opinion the "best" classifications are those which attempt to place organisms in groups on the basis of their phylogenetic relationships (presumed or otherwise). Admittedly this is not an easy task when dealing with the bacteria for which there is essentially no fossil record. It may not be possible to force them into structurally rigid classification systems similar to those of morphologically more complex organisms, but the basic philosophy of classification still applies. As Stanier and Van Niel[2] have pointed out, an imperfect natural system is preferable to a purely empirical one.

An overall view will show that microorganisms do fall into natural units which are recognizable and definable. What is not always appreciated is that these natural units cannot be defined in exactly the same way that species of higher organisms are defined,[3] but they can be described in similar terms. There are natural units (species) that have geographical–distribution area. For example, *Franciscella tularensis* occurs only in the North Temperate Zone, *Rickettsia tsutsugamushi* is restricted to parts of Asia, and *Bartonella bacilliformis* occurs only in Peru, Ecuador, and Colombia. If another dimension of distribution is added, namely, the specific habitat or ecosystem occupied by the organism, other natural units (on the species level as well as on the generic and higher taxonomic levels) can be recognized. The rickettsiae are good examples; certain "natural groups" are parasites or symbiotes of ticks, others of insects, and these may also have restricted geographic distribution coinciding with that of their hosts.

These natural groups are self-perpetuating. The rate of change (evolution) of bacterial "species" may appear to be greater than that of more complex organisms, but this is not necessarily true of "field populations". In the laboratory, "evolution" can be accelerated, but natural populations tend to change more slowly. *Yersinia pestis,* for example, does not appear to have changed substantially in the last 2000 years, at least with regard to its pathogenicity for man.

Morphologically distinguishable groups of microorganisms obviously exist, but morphological characters per se may not be as important in microbial taxonomy as they are for more complex organisms.[4] The addition of cytological, chemical, physical, and biochemical parameters to structural characteristics, however, does provide sufficient diversity and complexity to serve as useful criteria for group definition.[5] The relative proportion of nucleotide bases in microbial nucleic acid and DNA homology between organisms, as well as polypeptide fingerprinting, are becoming powerful tools for the systematist. Comparisons of macromolecular sequences can reveal evolutionary trends and phylogenetic relationships which may or may not fit with those derived by other methods.[6–9]

Microbial genetics, differing in many respects from that of higher organisms, has been used to some extent in systems of classification. Bacteria are essentially asexual, but gene-flow does occur through nonsexual mechanisms such as transformation, transduction, and perhaps others.[10,11]

A natural system of classification must take into account (as any good system should) all the parameters that can be measured and defined. Most important of all, natural units must be defined on the basis of all the characteristics of *natural populations,* not on those of laboratory stock cultures, which may or may not be representative of the population as a whole. A good deal of logic and reason, and not a little speculation, are also essential ingredients for devising a classification based on phylogenetic relationships. Excellent discussions of microbial classification and comprehensive bibliographies have been published[1,11–14] and will not be repeated here.

II. EVOLUTION OF RICKETTSIAL CLASSIFICATION

A. Arthropod Association and Pathogenicity

Wolbach,[15] recognizing the similarity of various rickettsia-like microorganisms that occur in insects and acarines, proposed that they be separated into three groups on the basis of their mode of life and pathogenicity for man. Thus, he placed all the nonpathogenic symbiotes of insects in one group, the obligate intracellular organisms pathogenic for man in another, and those that grow extracellularly in the gut lumen of lice or sheep keds (*Rickettsia* [=*Rochalimaea*] *quintana* and *Wolbachia melophagi*) in a third group. He also recognized that the extracellular rickettsiae closely resemble ordinary bacteria, the intracellular parasites differing from them in degree as a result of their specialization to an intracellular environment. These remarkably astute observations made over half a century ago are now generally accepted.

Wolbach was interested more in the organisms themselves than in the diseases they produced. About the same time, others were concerned with the diseases caused by these organisms in man. Megaw[16,17] was the first to bring a semblance of order to the perplexing array of exanthematous fevers recognized at the time. In this early stage, he grouped typhus, Rocky Mountain spotted fever, tsutsugamushi fever, and related diseases of uncertain origin into a single typhus group of fevers caused by rickettsiae and transmitted by arthropods. A short time later he also classified these typhus fevers according to vectors,[18] that is, louseborne, tickborne, and miteborne diseases; fleas were soon added to the list as vectors of fleaborne (endemic or murine) typhus.

Meanwhile Fletcher and Lesslar[19] were trying to unravel the puzzling array of typhus-like fevers occurring in Malaya. They made the extremely important observation that the sera of patients who recovered from some, but not all, types of typhus-like fevers in Malaya agglutinated certain strains of *proteus* bacilli. Audy and Savoor,[20] in the excellent history of the Institute for Medical Research in Kuala Lumpur, give a detailed account of this fascinating story which resulted in a classification of rickettsial diseases on the basis of the Weil-Felix reaction.

The Weil-Felix reaction is so intimately associated with the early typhus fever studies in Malaya that the story is worth repeating here. Professor W. J. Wilson was the first to discover that an organism isolated from the urine of a typhus patient was agglutinated by serum from other typhus patients. His observations were made in 1909. It wasn't until 1915 that Weil and Felix isolated *Bacillus proteus* (*Proteus vulgaris*) from the urine of a typhus patient and found that it was agglutinated by the patient's serum and that of other typhus patients. A year later Felix isolated another *proteus* bacillus, which he called the "X 19" strain, and it became the basis of the Weil-Felix reaction, which, to be fair, should be called the Wilson-Weil-Felix reaction.

In 1924, a sample of *B. proteus* X 19 was obtained from the Bland-Sutton Institute by A. N. Kingsbury and brought to the Institute for Medical Research in Kuala Lumpur, Malaya. This strain had been sent from the National Collection of Type Cultures to the Bland-Sutton Institute in 1921. Fletcher and Lesslar began to use the strain on so-called tropical typhus cases and found that serum from these patients agglutinated the organism. This linked the Malayan cases to the louseborne typhus fevers of Europe (fleaborne typhus had not yet been discovered), but the epidemiology of Malayan typhus was so different that suspicions were aroused. A detailed comparison of nine different *B. proteus* strains was then made.[21] Eight were similar to each other, the ninth was different. It was the "Kingsbury" strain. It was agglutinated by the serum of "typhus" cases

Table 1
CLASSIFICATION OF THE TYPHUS GROUP BASED ON THE *PROTEUS* OX AGGLUTINATION TEST

	Group 1	Group 2	Group 3
Proteus agglutinin	(OX 19)	(OX K)	(OX 19, OX K, OX 2—slight agglutination)
Human disease	Epidemic typhus, tabardillo, Brill's disease, endemic typhus	Tsutsugamushi fever, scrub typhus	Rocky Mountain spotted fever, Sao Paulo fever, *Fievre boutonneuse*, South African, Indian, Kenyan tick bite fever
Rickettsia	*Rickettsia prowazekii* *R. typhi*	*R. tsutsugamushi*	*R. rickettsii* *R. conorii*

which did not agglutinate any of the other strains and vice versa. It also did not produce indole by fermentation as did the other strains. Several strains that were unable to produce indole were compared to the Kingsbury strain, but they were not the same. The Kingsbury or "K" strain was unique. Subsequent use of these two strains of *Proteus vulgaris* resulted in the identification of two forms of typhus, one of which, scrub typhus or tsutsugamushi disease (chiggerborne rickettsiosis), was unique to Asia.[22] The other form was fleaborne typhus, which resembles louseborne typhus in possession of antigens similar to those of *proteus* OX 19, as it is generally referred to today.

When and how the National Collection of Type Cultures strain of *P. vulgaris* OX 19 got replaced by or mutated into *P. vulgaris* OX K will never be known. This fortuitous accident was instrumental in clarifying a very confusing situation in which two clinically similar but epidemiologically distinct diseases were found to be caused by different organisms transmitted by different vectors.

B. Immunological Characteristics

Classification of parasitic bacteria and viruses at the species level is based largely on antigenic relationships as measured by various serological techniques. Thus, Felix[23,24] distinguished three subgroups among the "typhus group" of fevers on the basis of the *Proteus* agglutination test (Table 1). This classification was for the epidemiologist and fit in well with Megaw's classification of the typhus fevers according to vectors. Those diseases transmitted by lice and fleas are in Group 1, those transmitted by trombiculid mites in Group 2, and those transmitted by ticks in Group 3. All subsequent workers have confirmed the integrity of these groups of rickettsiae, and, by working with the agents themselves, have further characterized the "species" of rickettsiae within each group and added new ones.

The development of immunological and serological techniques and their application to rickettsiae has further defined the species, and the process is still going on. The named *Rickettsia* species grouped according to modern immunological tests are shown on Table 2.

1. The Louseborne and Fleaborne (OX 19-Agglutinating) Group

The agents of louseborne typhus (*R. prowazekii*) and fleaborne typhus (*R. typhi*) were long known to be closely related on the basis of many observations, not the least of which was their ability to produce cross-immunity in guinea pigs.[25] Crude antigens also produced cross-reactions in the complement fixation test, but Plotz[26] demonstrated that

Table 2
IMMUNOLOGICAL GROUPING OF THE NAMED RICKETTSIA SPECIES

Group 1	Group 2	Group 3
Rickettsia prowazekii	*Rickettsia tsutsugamushi*	*Rickettsia rickettsii*
R. typhi		*R. sibirica*
R. canada		*R. conorii*
		R. parkeri
		R. akari
		R. australis
		R. montana
		R. rhipicephali
		R. slovaca

purified antigens are highly specific and can distinguish between them. Gildemeister and Haagen[27] added another tool to the serologists' armamentarium by showing that murine typhus rickettsiae produce a toxic factor with immunologic properties, and Hamilton[28] demonstrated that toxins produced by *R. typhi* and *R. prowazekii* are immunologically distinct. The results obtained with the toxin neutralization test, the conclusions drawn from the complement fixation reaction, and many other lines of evidence indicated that *R. prowazekii* and *R. typhi* are closely related but antigenically distinct organisms. Furthermore, immunization and cross-challenge tests showed a weak relationship between this group and the tickborne group. Guinea pigs immunized with typhus rickettsiae are often also protected against challenge with virulent Rocky Mountain spotted fever rickettsiae.[29]

A Tickborne Member of the Typhus Group—In 1963 strains of a *R. typhi*-like rickettsia were isolated from the rabbit tick *Haemaphysalis leporispalustris*, taken from a sentinel rabbit and from a wild snowshoe hare in Ontario, Canada.[30] In complement fixation tests with guinea pig sera, this agent was shown to be nearly identical to *R. typhi* rather than to any known tickborne agent. It can be distinguished from *R. typhi* and *R. prowazekii*, however, by toxin neutralization tests and by the species-specific complement fixation test obtained with mouse antisera, which had proved useful in differentiating other closely related rickettsiae.[31] This new rickettsia, named *R. canada*, appears to belong in the typhus group on the basis of antigenic relationships even though it is apparently tickborne rather than carried by fleas or lice.

2. The Trombiculid Miteborne (OX K-Agglutinating) Group

Rickettsia tsutsugamushi is immunologically distinct from all other rickettsiae but shows considerable antigenic variation within the species. Three major antigenic types are generally recognized: Karp, Kato, and Gilliam,[32] but the recent work of Elisberg et al.[33] and others suggests that other antigenic variants occur naturally in Thailand and elsewhere. Much more work is needed before the highly complex antigenic properties within this group can be resolved. The uniqueness of *R. tsutsugamushi* among the Rickettsieae, however, is unquestioned.

3. The Tickborne "Spotted Fever" Group

The spotted fever (tickborne) rickettsiae form a heterogeneous group that can be defined immunologically on the basis of a common antigenic component. Lackman and Pickens[34] defined the group as one containing all rickettsiae having a soluble antigen which fixes complement in the presence of serum of guinea pigs convalescent from

Rocky Mountain spotted fever. The individual species can be distinguished by their reactions in one or more of the following tests:

1. Complement fixation, using washed rickettsial suspensions reacted with antisera prepared in guinea pigs inoculated with living rickettsial suspensions[35]
2. Cross-challenge, in guinea pigs immunized with either living or formalin-killed rickettsial suspensions[36,37]
3. Toxin neutralization[38]
4. Mouse protection against a toxic challenge[39]
5. Complement fixation using mice for the production of species-specific CF antibodies.[31]

Lackman et al.[37] clearly summarized the serological and immunological relationships between these rickettsiae and separated them into four distinct but related subgroups. *Rickettsia rickettsii* and *R. sibirica* are in subgroup A and *R. conorii* and *R. parkeri* in subgroup B on the basis of strong cross-reactions in the toxin neutralization test. The relationship between subgroups A and B is indicated by a slight cross-reaction between *R. sibirica* and *R. conorii* in the CF test. The various geographical strains of *R. conorii* (South African tick-bite fever, Kenya tick typhus, and Indian tick typhus) are antigenically identical to the agent of *fievre boutonneuse*.[38] Subgroup C consists of *R. australis* and *R. akari* whose relationship is shown by the guinea pig cross-challenge test. Subgroup D contains *R. montana* and Western Montana U agent, an avirulent strain similar to *R. rickettsii*, on the basis of CF antibody studies in mice. More recent immunological groupings will be discussed in Volume I, Chapter 5.

C. The *Proteus* Nonagglutinating Organisms

Coxiella burnetii and *Rochalimaea quintana* are antigenically distinct from all recognized species of *Rickettsia*. Exhaustive comparative serological and immunological studies have not been reported, but Vinson and Campbell[40] have shown that *R. quintana* and *C. burnetii* do not cross-react in the complement fixation test, nor does *R. quintana* immune rabbit serum contain antibodies to soluble *R. prowazekii* antigen. These two organisms appear to be as distinct antigenically from the main *Rickettsia* stem as they are distinct from it in many other characteristics. They will be discussed in greater detail in Volume II, Chapters 10 and 11.

III. PHYLOGENETIC RELATIONSHIPS AND CLASSIFICATION ACCORDING TO VECTOR

Based on immunological criteria the rickettsia stem, divorced of *C. burnetii* and *R. quintana*, is composed of a rather homogeneous group of organisms (Table 2). Any natural classification will have to include the agents of typhus fevers (Group 1) and the tickborne agents (Group 3). The members of these antigenic groups are closely related phylogenetically and the phylogenetic relationship of *R. tsutsugamushi* to Groups 1 and 3 can be hypothesized, but may be rather remote. The relationships of the other "rickettsial" lines are less easily derived. These relationships as well as those of other families in the order will be explored in some detail in the following chapters.

A. Terminology

A variety of terms commonly used to refer to rickettsiae and rickettsial diseases are sometimes used inaccurately, and their meaning often is not entirely clear. Common usage of certain terms over many years has made them household words in various parts of the world, and they will continue to be used regardless of the fact that they may be etymologically incorrect. Generally, two sets of terms are necessary to clearly distinguish between the organism and the disease it causes in man. Specific terms to refer to individual organisms are also required in some instances. The context in which terms are used often determines their exact meaning, but this is not always the case. In the following discussions, I have attempted to minimize the confusion by sticking wherever possible to a standard set of terms to refer to the rickettsiae and the diseases they cause. In some cases, terms that are widely used and whose meanings are clear may be used interchangeably if there is no chance for confusion. In general, the terminology suggested by Traub and Wisseman[41] will be used throughout. This terminology recognizes the fact that infection is not always (perhaps seldom) associated with disease. Thus, infection of whatever nature is referred to as a "rickettsiosis." The type of rickettsiosis is indicated by prefixing the name of the common vector or arthropod maintaining host when relevant or known. When referring to the organisms themselves as a group, the term rickettsia (with a small "r") with the appropriate prefix is employed.

B. Tickborne Rickettsiae

All the organisms which are maintained naturally by ticks are included in the term tickborne rickettsiae, (singular, rickettsia). The most common synonyms for this group are tick typhus rickettsiae or spotted fever rickettsiae. The organisms in this group and the diseases they cause in man are

Rickettsia conorii	*Fievre boutonneuse*
R. sibirica	North Asian (Siberian) tick typhus
R. rickettsii	Rocky Mountain spotted fever, spotted fever
R. australis	Queensland tick typhus
R. akari	Rickettsialpox
R. canada	No disease known
R. parkeri	No disease known
R. montana	No disease known
R. rhipicephali	No disease known
R. slovaca	No disease known

This classification suffers from the same problem inherent in all generalizations. Lumped together are organisms which can be readily separated into distinct categories of their own on the basis of other biological criteria. *Rickettsia canada,* although tickborne, is antigenically related to the insectborne typhus group rather than to the spotted fever or tick typhus group. Also, *R. akari* technically is not tickborne, but is maintained naturally by a mesostigmatic mite even though it is antigenically and biologically close to the spotted fever group. The mesostigmata (mites), however, are closely related phylogenetically to the Metastigmata (ticks) and in many respects, they resemble tiny ticks. Since *R. akari* is the only rickettsia known to be associated with mesostigmatic mites, and because of its other biological properties, it is included in the tickborne rickettsiae.

The Wolbachieae are symbiotes of ticks and other arthropods and they do not infect vertebrates as far as is known. The tickborne members of this group may be phylogenetically related to the tickborne rickettsiae, but there is reason to believe that the insectborne Wolbachieae are phylogenetically unrelated (Volume II, Chapter 8).

The association of members of the Ehrlichieae with arthropods is not so well established as for the Rickettsieae and Wolbachieae, and phylogenetic relationships are tenuous at best. In one species, *Neorickettsia helminthoeca*, the natural invertebrate host appears to be a trematode worm, not an arthropod at all (see Volume II, Chapter 9).

In the family Anaplasmataceae, there are organisms whose association with ticks is reasonably well-established, others for which it is debatable, and still others for which it doesn't exist at all. The phylogenetic relationship of the anaplasmas with the rickettsiae may be very remote at best (see Volume II, Chapter 11).

Coxiella burnetii, although it may be tickborne, is not included in this group for reasons that will be discussed in Volume II, Chapter 10.

C. Insectborne Rickettsiae

The rickettsiae maintained in or closely associated with insects are classed as the insectborne rickettsiae and include *R. typhi*, fleaborne (endemic or murine) typhus, and *R. prowazekii*, louseborne (epidemic or classical) typhus. *Rochalimaea quintana*, the etiologic agent of trench fever in man, might logically be included in this group since its natural host is the human body louse, but there are important reasons to doubt its membership in the Rickettsiales (Volume II, Chapter 11).

One of the three species in the small family Bartonellaceae, *Bartonella bacilliformis*, is biologically associated with phlebotomine flies and thus may be considered to be arthropodborne. The other two species are not associated with arthropods at all. All three species grow well on nonliving bacterial media and in many other ways are not rickettsial in nature (see Volume II, Chapter 11).

D. Trombiculidborne Rickettsiae

The rickettsiae maintained by trombiculid mites, the numerous serotypes of *R. tsutsugamushi*, are classified as trombiculidborne rickettsiae. The infection they cause is termed chiggerborne rickettsiosis in recognition that only the larvae (chiggers) of the trombiculid mites are parasitic on vertebrates, and only they can transmit the rickettsiae (see Volume II, Chapter 7).

IV. NATURAL CLASSIFICATION OF THE RICKETTSIAE

Development of a natural or phylogenetic classification of the rickettsiae is not an impossible or impractical task. The obligate association of most members of the group with vertebrates and arthropods suggests a long-term, perhaps evolutionary, relationship so that the origin, development, and divergence of the rickettsiae may be traced through the origin, development, and divergence of their arthropod and vertebrate hosts. This approach is utilized in the present review and discussion of rickettsial ecology and evolution.

Documentation of facts and statements is provided whenever possible by citing the original references. Most of the Russian literature cited was seen as translations through the generosity of Dr. Harry Hoogstraal, U.S. Navy Medical Research Unit No. 3, Cairo, Egypt. Considerable use also was made of Hoogstraal's multivolume *Bibliography of Ticks and Tickborne Diseases*[42] and the *Index-Catalogue of Medical and Veterinary Zoology* complied by Doss et al.[43]

REFERENCES

1. **Cowan, S. T.,** The principles of microbial classification. Introduction, *J. Gen. Microbiol.*, 12, 314, 1955.
2. **Stanier, R. Y. and Van Niel, C. B.,** The main outlines of bacterial classification, *J. Bacteriol.*, 42, 437, 1941.
3. **Huxley, J.,** Introductory: towards the new systematics, in *The New Systematics,* Huxley, J., Ed., Clarendon Press, Oxford, 1940, 1.
4. **Gibson, T.,** The principles of microbial classification. General morphology, *J. Gen. Microbiol.*, 12, 324, 1955.
5. **Bisset, K. A.,** The principles of microbial classification. The value of cytological studies in elucidating natural relationships among bacteria, *J. Gen. Microbiol.*, 12, 325, 1955.
6. **Zuckerkandl, E. and Pauling, L.,** Molecules as documents of evolutionary history, *J. Theor. Biol.*, 8, 357, 1965.
7. **Wilson, A. C., Carlson, S. S., and White, T. J.,** Biochemical evolution, *Annu. Rev. Biochem.*, 46, 573, 1977.
8. **Stanier, R. Y., Wachter, D., Gasser, C., and Wilson, A. C.,** Comparative immunological studies of two *Pseudomonas* enzymes, *J. Bacteriol.*, 102, 351, 1970.
9. **Ambler, R. P., Meyer, T. E., and Kame, M. D.,** Primary structure determination of two cytochromes C_2: close similarity to functionally unrelated mitrochondrial cytochrome C, *Proc. Natl. Acad. Sci. U.S.A.*, 73, 472, 1976.
10. **Pontecorvo, G.,** The principles of microbial classification. The impact of genetics, *J. Gen. Microbiol.*, 12, 330, 1955.
11. **Jones, D. and Sneath, P. H. A.,** Genetic transfer and bacterial taxonomy, *Bacteriol. Rev.*, 34, 40, 1970.
12. **Stanier, R. Y. and Van Niel, C. B.,** The concept of a bacterium, *Arch. Mikrobiol.*, 42, 17, 1962.
13. **Floodgate, G. D.,** Some remarks on the theoretical aspects of bacterial taxonomy, *Bacteriol. Rev.*, 26, 277, 1962.
14. **Gibbons, N. E. and Murray, R. G. E.,** Proposals concerning the higher taxa of bacteria, *Int. J. Syst. Bacteriol.*, 28, 1, 1978.
15. **Wolbach, S. B.,** The rickettsia and their relationship to disease, *JAMA*, 84, 723, 1925.
16. **Megaw, J. W. D.,** A typhus-like fever in India, possibly transmitted by ticks, *Indian Med. Gaz.*, 56, 361, 1921.
17. **Megaw, J. W. D.,** The typhus group of fevers, *Indian Med. Gaz.*, 59, 68, 1924.
18. **Megaw, J. W. D.,** Indian tick typhus, *Indian Med. Gaz.*, 60, 58, 1925.
19. **Fletcher, W. and Lesslar, J. E.,** Tropical typhus in the Federated Malay States, *Bull. Inst. Med. Res. F.M.S.*, 2, 1, 1925.
20. **Audy, J. R. and Savoor, S. R.,** Typhus, in *The Institute for Medical Research. 1900–1950,* Field, J. W. and Byron, F. E., Eds., The Government Press, Kuala Lumpur, Malaya, 1951, 184.
21. **Fletcher, W. and Lesslar, J. E.,** A comparison of some strains of *Bacillus proteus* employed in the Weil-Felix reaction, *Trans. Far-East Assoc. Trop. Med.*, 2, 775, 1925.
22. **Fletcher, W. and Lessler, J. E.,** The Weil-Felix reaction in sporadic tropical typhus, *Bull. Inst. Med. Res. F.M.S.*, 1, 1, 1926.
23. **Felix, A.,** The serology of the typhus group of diseases, *Trans. Soc. Trop. Med. Hyg.*, 29, 113, 1935.
24. **Felix, A.,** The typhus group of fevers. Classification, laboratory diagnosis, prophylactic inoculations, and specific serum treatment. *Brit. Med. J.*, 2, 597, 1942.
25. **Zinsser, H.,** Epidemiology and immunity in the rickettsial diseases. Harvard Symposium, in *Viruses and Rickettsial Diseases,* Harvard University Press, Cambridge, Mass., 1941, 872.
26. **Plotz, H.,** Complement fixation in rickettsial diseases, *Science,* 97, 20, 1943.
27. **Gildemeister, E. and Haagen, E.,** Fleckfieber-studien. I. Mitteilung: Nachweis eins Toxins in Rickettsien-Erkultwen *(Rickettsia mooseri), Deutsch Med. Wochenschr.*, 661, 878, 1940.
28. **Hamilton, H. L.,** Specificity of the toxic factors associated with the epidemic and the murine strains of typhus rickettsiae, *Am. J. Trop. Med.*, 25, 391, 1945.
29. **Parker, R. R., Pickens, E. G., Lackman, D. B., Bell, E. J., and Thrailkill, F. B.,** Isolation and characterization of Rocky Mountain spotted fever rickettsiae from the rabbit tick *Haemaphysalis leporus-palustris* Packard, *Public Health Rep.*, 66, 455, 1951.
30. **McKeil, J. A., Bell, E. J., and Lackman, D. B.,** *Rickettsia canada:* a new member of the typhus group of rickettsiae isolated from *Haemaphysalis leporus-palustris* ticks in Canada, *Can. J. Microbiol.*, 13, 503, 1967.

31. **Pickens, E. G., Bell, E. J., Lackman, D. B., and Burgdorfer, W.,** Use of mouse serum in identification and serologic classification of *Rickettsia akari* and *Rickettsia australis, J. Immunol.*, 94, 883, 1965.
32. **Shishido, A.,** Strain variation of *Rickettsia orientalis* in the complement fixation test, *Jpn. J. Med. Sci. Biol.*, 17, 59, 1964.
33. **Elisberg, B. L., Sankasuvana, V., Campbell, J. M., Bozeman, F. M., Bodhidatta, P., and Rapmund, G.,** Physiographic distribution of scrub typhus in Thialand, *Acta Med. Biol.*, 15 (Suppl.), 61, 1967.
34. **Lackman, D. B. and Pickens, E. G.,** Antigenic types in the Rocky Mountain spotted fever group of rickettsiae, *Bacteriol. Proc.*, 3, 219, 1953.
35. **Plotz, H., Bennett, B. L., Wertman, K., and Snyder, M. J.,** Cross-reacting antibodies in Rocky Mountain spotted fever, *Proc. Soc. Exp. Biol. Med.*, 57, 336, 1944.
36. **Davis, G. E. and Parker, R. R.,** Comparative experiments in spotted fever and boutonneuse fever, *Public Health Rep.*, 49, 423, 1934.
37. **Lackman, D. B., Bell, E. J., Stoenner, H. G., and Pickens, E. G.,** The Rocky Mountain spotted fever group of rickettsias, *Health Lab. Sci.*, 2, 135, 1965.
38. **Bell, E. J. and Stoenner, H. G.,** Immunologic relationships among the spotted fever group of rickettsias determined by toxin neutralization tests in mice with convalescent animal serums, *J. Immunol.*, 84, 171, 1960.
39. **Bell, E. J. and Stoenner, H. B.,** Spotted fever vaccine; potency assay by direct challenge of vaccinated mice with toxin of *Rickettsia rickettsii, J. Immunol.*, 84, 737, 1961.
40. **Vinson, J. W. and Campbell, E. S.,** Complement fixing antigens from *Rickettsia quintana, Acta Virol. Engl. Ed.*, 12, 54, 1968.
41. **Traub, R. and Wisseman, C. L.,** The ecology of chigger-borne rickettsiosis (Scrub typhus), *J. Med. Entomol.*, 11, 237, 1974.
42. **Hoogstraal, H.,** Bibliography of Ticks and Tick-borne Diseases, Vols. 1–5, Special Publication, U.S. Naval Med. Res. Unit No. 3, Cairo, 1970—1978.
43. **Doss, M. A., Farr, M. N., Roach, K. F., and Anastos, G.,** Index-Catalogue of Medical and Veterinary Zoology, Ticks and Tickborne Diseases, Vol. 1 and 2, U.S. Department of Agriculture, Washington, D.C. 1974.

Chapter 2

THE DISTRIBUTION AND RELATIONSHIPS OF RICKETTSIAE

N. Marchette

TABLE OF CONTENTS

I. The Present System of Classification 12
A. Separation of Rickettsiae and Viruses 12
B. Differentiation from Other Bacteria 12
C. Chlamydiae 14

II. Taxonomy and Distribution 14
A. Rickettsiaceae 14
1. Tribe Rickettsieae 16
a. Tickborne Group 16
b. Insectborne Group 18
c. Trombiculidborne Group 18
d. Summary 18
2. Tribe Ehrlichieae 18
a. *Ehrlichia* 18
b. *Cowdria* 18
c. *Neorickettsia* 20
3. Tribe Wolbachieae 20
a. *Wolbachia* 20
b. *Symbiotes*, *Blattabacterium*, and *Rickettsiella* 20
B. Bartonellaceae 21
C. Anaplasmataceae 21
1. *Anaplasma* and *Paranaplasma* 21
2. *Aegyptianella* 21
D. Chlamydiales 22

III. Definition of the Rickettsiaceae 22

IV. Origin and Evolution of the Rickettsiae 23

References 25

"What is now proved was once only imagined."

William Blake

I. THE PRESENT SYSTEM OF CLASSIFICATION

The Seventh Edition of *Bergey's Manual of Determinative Bacteriology* (henceforth referred to as *Bergey's Manual*)[1] separated the rickettsiae from the blue-green algae and bacteria and placed them in a new class (Microtatobiotes) with the viruses (Table 1). This was a legitimate and authoritative attempt to establish a category for those organisms which, as a group, differ superficially at least from other microorganisms traditionally accepted and referred to as bacteria. It was a classification of convenience, more pragmatic than natural, and one that survived only until the Eighth Edition of *Bergey's Manual* was prepared.

A. Separation of Rickettsiae and Viruses

As knowledge accumulated, it became more and more apparent that the Order Rickettsiales resembles certain orders of the Schizomycetes (bacteria) more closely than it does the Order Virales and that the two groups are not at all related in a phylogenetic sense. The reasons given by Philip[2] for establishing a higher taxon (Class Microtatobiotes) to include both rickettsiae and viruses are not convincing. Many investigations have established the unequivocal bacterial nature of the Rickettsiales and the unique qualities of the Virales. The viruses lack enzyme systems, cell walls, energy systems, and cytoplasm, are insensitive to antibiotics, do not multiply by binary fission, and possess a single type of nucleic acid.[3] The rickettsiae (broadly defined) may lack certain enzymes, and some are deficient in energy-producing mechanisms, but they are all generally inhibited by chemotherapeutic agents, possess cell membranes, multiply by binary fission,[4-6] contain both ribose nucleic acid and deoxyribose nucleic acid,[3] and have a capacity for incorporating amino acids into proteins which is blocked by chloramphenicol.[7] Thus, there are compelling reasons for placing the Virales, which possess none of these characteristics, in a separate taxon in recognition of their unique qualities. The Eighth Edition of *Bergey's Manual*[8] corrects this situation by dropping the Virales and including the rickettsiae and chlamydeae in the orders Rickettsiales and Chlamydiales, respectively, in a separate class of obligate intracellular scotobacteria parasitic in eukaryotes (Table 2). The mycoplasmas, some of which may be associated with arthropods, are placed in a separate class, primarily on the basis of their lack of a cell wall.

B. Differentiation from Other Bacteria

The Rickettsiales and Chlamydiales possess most, if not all, of the characteristics attributable to other scotobacteria. They can be differentiated from them by being intracellular parasites of eukaryotic cells and from the Mycoplasmatales by possessing cell walls (see Volume II, Chapter 11 for discussion of the systematic position of the Anaplasmataceae). Thus, the Rickettsiales are now back in the nonphototropic bacteria (scotobacteria) and the viruses are a group by themselves as Levaditi et al.[9] (1964) had earlier proposed. These authors further arranged the Rickettsiaceae, Chlamydiaceae, and Bartonellaceae as families within one superfamily of the Order. This is still a completely artificial classification, lumping together groups of organisms that may resemble each other in superficial ways but that are probably completely unrelated phylogenetically. The same criticism applies to Krieg's[10] Class Rickettsoideae, with the Order Rickettsiales comprising the two families Chlamydozoaceae and Rickettsiaceae. This

Table 1
CLASSIFICATION OF THE RICKETTSIAE IN THE PROTOPHYTA ACCORDING TO *BERGEY'S MANUAL OF DETERMINATIVE BACTERIOLOGY*, 7th EDITION[1]

Division I. Protophyta
- Class I. Schizophyceae Cohn—Blue-green algae
- Class II. Schizomycetes von Naegeli—Bacteria, including Mycoplasmas
- Class III. Microtatobiotes Philip—Rickettsiae and Viruses
  - Order I. Rickettsiales Buchanan and Buchanan emend. Gieszczykiewicz
    - Family I. Rickettsiaceae Pinkerton
      - Tribe I. Rickettsieae Philip
      - Tribe II. Ehrlichieae Philip
      - Tribe III. Wolbachieae Philip
    - Family II. Chlamydiaceae Rake
    - Family III. Bartonellaceae Gieszczykiewicz
    - Family IV. Anaplasmataceae Yakimov
  - Order II. Virales Breed, Murray and Hitchens

Table 2
CLASSIFICATION OF THE RICKETTSIAE AND CHLAMYDIAE IN THE 8th EDITION OF *BERGEY'S MANUAL*[8]

Kingdom Procaryotae

Division I: The Cyanobacteria. Phototrophic procaryotes ("Photobacteria") including blue-green algae

Division II: The Bacteria. Procaryotes indifferent to light ("Scotobacteria")
- Class I. The Bacteria
- Class II. Obligate intracellular scotobacteria in eukaryotic cells—includes Rickettsiae and Chlamydiae
  - Part 18. The Rickettsias
    - Order I. Rickettsiales Gieszczkiewicz
      - Family 1. Rickettsiaceae Pinkerton
      - Family 2. Bartonellaceae Gieszczkiewicz
      - Family 3. Anaplasmataceae Philip
    - Order II. Chlamydiales Storz and Page
- Class III. Mollicutes Edward and Freundt. Scotobacteria without cell walls
  - Part 19. The Mycoplasmas
    - Order I. Mycoplasmatales Freundt

was partially corrected by separation of the chlamydiae into a separate order (Table 2).

The true bacterial nature of rickettsiae and chlamydiae is well documented in a number of biochemical and morphological studies.[11-15] In the first study of rickettsiae with the electron microscope, Plotz et al.[16] discovered that they contain dense granules embedded in a protoplasmic substance enclosed within a limiting membrane. This was later confirmed by Weiss[17] and by Mudd and Anderson,[18] who referred to the limit membrane as a "limiting cell wall clearly distinct from the inner protoplasm." Recent studies utilizing refined techniques and more powerful electron microscopes with higher resolving power have shown the complex nature of the cell wall. Thus Shkolnik et al.[19] found that the cell wall of *Rickettsia prowazekii* consists of two membranes, the inner one a cytoplasmic membrane and the outer one a unit membrane analogous to a bacterial cell wall. The rickettsial cell wall appeared structurally less complex that that of some bacteria,[20–22] but a physically simpler structure may be an adaptation better suited to the intracellular environment. Later work, however, suggests that the cell wall of *R. prowazekii* is not trilamellar as previously thought, but actually consists of five layers, making it at least as complex as that of other Gram-negative bacteria.[23–26] Between the cytoplasmic membrane and the cell wall of *Rickettsia* spp. is a peptidoglycan layer, a structure not unlike that of *Escherichia coli*.[27] The presence of a slime layer of capsular material external to the cell wall of *R. prowazekii* and *R. rickettsii* also has been demonstrated.[28] The rickettsial cell wall is chemically similar to its bacterial analogue in containing amino acids, oligosaccharides, and muramic acid—a substance specific to the mucopeptides of bacterial cell walls.[29–32] In reviewing the biochemical and biological similarities between rickettsiae and bacteria, Wisseman[33] concluded that the evidence overwhelmingly supports the view that rickettsiae constitute a special group of bacteria.

C. Chlamydiae

The chlamydial cell wall also resembles that of Gram-negative bacteria,[34–39] but there are some differences,[38,39] chief among them being the apparent absence or reduced amount of mucopeptide.[40] These organisms also have a complex life cycle in which one form, the reticulate body, resembles bacterial L-forms or cell wall defective variants. Kordova[41] considers that these obligate intracellular, cell wall defective organisms with complex stages of development may be phylogenetically intermediate between "L" forms of some extinct free-living bacteria and true large viruses. This hypothesis is no longer as widely accepted as it was at one time.

Before the invention and application of the electron microscope, it was not clear that the so-called larger viruses and even the rickettsiae were actually cellular organisms possessing complex membranes, cytoplasmic structures, and nuclear material. It was apparent, however, that "rickettsiae," although intracellular parasites of minute size, were capable of a certain amount of ". . . independent metabolic activity within their host cells,"[42] so it was not unusual to consider them to be intermediate between certain pathogenic Gram-negative bacteria and viruses.

The organisms now placed in the Chlamydiales originally were considered by some to be "large viruses", and viral terminology is still used to describe them. Intracellular colonies of organisms are referred to as inclusion bodies and the individual organisms in the inclusions as elementary bodies. They are actually more rickettsia-like than virus-like and would no doubt have been placed with the rickettsiae except that insect or acarine transmission could not be demonstrated. Insect transmission of the trachoma agent was suggested several decades ago,[43] but biological transmission was never proven. Mechanical transmission by flies and other insects is possible and may occur occasionally. Chlamydial agents have been isolated from acarines,[44–46] but to date attempts to demonstrate biological transmission by ticks and other arthropods have not been notably successful. A report of three cases of chlamydial infection after tick bite[47] has not been verified, and the data presented by the authors do not establish with any certainty that a *Chlamydia* caused the infections.

II. TAXONOMY AND DISTRIBUTION

A. Rickettsiaceae

The family Rickettsiaceae, as presently constituted, is divided into three tribes: Rickettsieae, Ehrlichieae, and Wolbachieae (Table 3). As an artifical system, this has proved to be a most useful classification, but organisms representing diverse phylogenetic lines still are lumped together.

Table 3
THE PRESENT CLASSIFICATION OF THE RICKETTSIALES (FROM *BERGEY'S MANUAL*[8] WITH ADDITION OF NEWLY DESCRIBED SPECIES)

Family I. Rickettsiaceae Pinkerton
- Tribe I. Rickettsiae Philip
  - Genus I. *Rickettsia* da Rocha-Lima
    1. *R. prowazekii* da Rocha Lima
    2. *R. typhi* (Wolbach and Todd) Philip
    3. *R. canada* McKiel, Bell, and Lackman
    4. *R. rickettsii* (Wolbach) Brumpt
    5. *R. sibirica* Zdrodovskii
    6. *R. conorii* Brumpt
    7. *R. australis* Philip
    8. *R. akari* Huebner, Jellison, and Pomerantz
    9. *R. parkeri* Lackman et al.
    10. *R. montana* Lackman, Bell, Stoenner, and Pickens
    11. *R. rhipicephali* Burgdorfer, Brinton, Krynski, and Philip
    12. *R. slovaca* Urvolgyi and Brezina
  - Genus II. *Rochalimaea* (Macchiavello) Krieg
    1. *R. quintana* (Schmincke) Kreig
  - Genus III. *Coxiella* (Philip) Philip
    1. *C. burnetii* (Derrick) Philip
- Tribe II. Ehrlichieae Philip
  - Genus IV. *Ehrlichia* Moshkovskiy
    1. *E. canis* (Donatien and Lestoquard) Moshkovskiy
    2. *E. phagocytophila* (Foggie) Philip
  - Genus V. *Cowdria* Moshkovskiy
    1. *C. ruminantium* (Cowdry) Moshkovskiy
  - Genus VI. *Neorickettsia* Philip, Hadlow, and Hughes
    1. *N. helminthoeca* Philip, Hadlow, and Hughes
- Tribe III. Wolbachieae Philip
  - Genus VII. *Wolbachia* Hertig
    1. *W. pipientis* Hertig
    2. *W. melophagi* (Nöller) Philip
    3. *W. persica* Suitor and Weiss
  - Genus VIII. *Symbiotes* Philip
    1. *S. lectularius* (Arkwright et al.) Philip
  - Genus IX. *Blattabacterium* Hollande and Favre
    1. *B. cuenoti* (Mercier) Hollande and Favre
  - Genus X. *Rickettsiella* Philip
    1. *R. popilliae* (Dutky and Gooden) Philip

Family II. Bartonellaceae
- Genus I. *Bartonella* Strong, Tyzzer, and Sellards
  1. *B. bacilliformis* (Strong et al.)
- Genus II. *Grahamella* Brumpt
  1. *G. talpae* Brumpt
  2. *G. peromysci* Tyzzer

Family III. Anaplasmataceae
- Genus I. *Anaplasma* Theiler
  1. *A. marginale* Theiler
  2. *A. ovis* Lestoquard
- Genus II. *Paranaplasma* Kreier and Ristic
  1. *P. caudatum* Kreier and Ristic
  2. *P. discoides* Kreier and Ristic
- Genus III. *Aegyptianella* Carpano
  1. *A. pullorum* Carpano
- Genus IV. *Haemobartonella* Tyzzer and Weinman
  1. *H. muris* (Mayer) Tyzzer and Weinman

Table 3 (continued)
THE PRESENT CLASSIFICATION OF THE RICKETTSIALES (FROM *BERGEY'S MANUAL*[8] WITH ADDITION OF NEWLY DESCRIBED SPECIES)

	2. *H. felis* (Clark) Flint and McKelvie
	3. *H. canis* (Kikuth) Tyzzer and Weinman
Genus	V. *Eperythrozoon* Schilling
	1. *E. coccoides* Schilling
	2. *E. ovis* Neitz, Alexander, and du Toit
	3. *E. suis* Splitter
	4. *E. parvum* Splitter
	5. *E. wenyoni* Adler and Ellenbogen

1. Tribe Rickettsieae

The presently constituted Tribe Rickettsieae includes the type genus *Rickettsia* and the monotypic genera *Coxiella* and *Rochalimaea*. Philip divided the genus *Rickettsia* into four subgenera (*Rickettsia, Zinssera, Dermacentroxenus* and *Rochalimaea,* but these have been dropped from the Eighth Edition of *Bergey's Manual*,[8] and *Rochalimaea* raised to generic status. Eventually the trombiculidborne rickettsiae may also be recognized as a separate genus, as I believe it should be. The tickborne and insectborne rickettsiae are perhaps too closely related to be placed in separate genera, although phylogenetically such a separation may be legitimate.

For reasons that will become clear in later chapters, *Rochalimaea quintana* and *Coxiella burnetii* cannot be included in the Tribe Rickettsieae and probably not even in the family Rickettsiaceae.

a. Tickborne Group

Rickettsia conorii is presently distributed throughout the Ethiopian Region and portions of the southwestern Palearctic. (The Faunal Regions of the world are shown in Figure 1). The tick typhus rickettsiae of the Oriental Region are probably distinct from *R. conorii* to the west and *R. sibirica* to the north. Strains isolated in West Pakistan and Thailand are antigenically unrelated to either. The only strains of tick typhus rickettsiae isolated in India have been identified as *R. conorii,* all from domestic animal ticks; but it is not known for certain if the indigenous form in native ticks, if such exists, is the Oriental or African species or a distinct organism.

Rickettsia sibirica is widely distributed in the Palearctic Region east of the Urals. There is little evidence at this time to indicate that distinct serotypes exist, but it would be surprising if they do not.

Rickettsia australis is confined to the Australo-Papuan Region. The more or less closely related, mesostigmatic mite-transmitted *R. akari* occurs along the eastern edge of the Palearctic and eastern edge of the Nearctic.

Rickettsia rickettsii is widely distributed throughout both continents in the New World.

Rickettsia parkeri is limited, as far as is known, to the southern U.S., but its range may be much wider, perhaps extending into Central America or even South America.

Rickettsia canada has been found only in North America.

Also included among the tick typhus group are a number of recently described serotypes in the U.S., Europe, and Asia. They are antigenically distinct from all the previously recognized species and some have been named. They will be discussed in Volume I, Chapter 5.

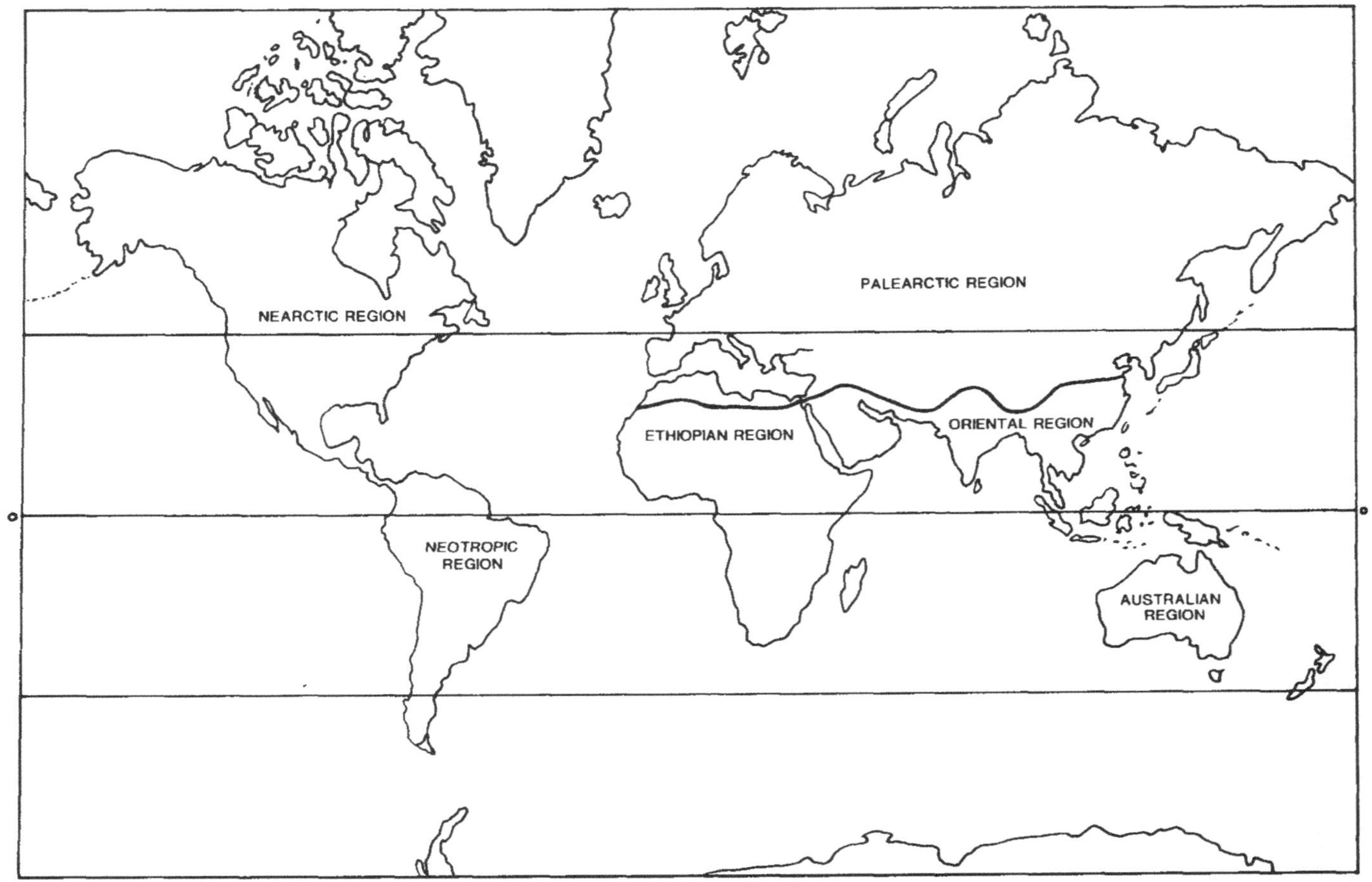

FIGURE 1. Faunal regions of the world.

b. Insectborne Group

Rickettsia typhi is worldwide in distribution, occurring wherever rats are found. *Rickettsia prowazekii* at one time also existed throughout the world wherever human beings lived. Extensive and largely successful control and eradication of human lice in recent decades have restricted the distribution of *R. prowazekii* to certain regions in Africa, South America, and Eurasia.

c. Trombiculidborne Group

Rickettsia tsutsugamushi is widely distributed in the Oriental Region, including India and Pakistan, and extends along the Indonesian Island chain into New Guinea and Australia. It is also found along the eastern portion of the Palearctic including mainland China, Russia, and the Japanese Archipelago. From the Oriental Region, it extends north into the mountain systems separating this tropical region from the temperate South Palearctic.

d. Summary

The various groups of tickborne rickettsiae thus occur in geographically distinct regions separated for the most part by natural boundaries (Figure 2). Their present geographic isolation, along with the demonstrably close antigenic and ecologic relationships within the group, suggest a common ancestor. This ancestral form may have arisen as a tick symbiote that became adapted to a temporary existence in mammals during part of its life history. This took place in the Old World Tropics—probably in what is now the Oriental Region—during the decline of the reptiles and the divergence and rapid development of the mammals. With the isolation of Africa from Asia and the development of *Rhipicephalus* ticks, the *R. conorii* branch developed in Africa and contributed no more to rickettsial evolution except perhaps as the recent ancestor of *R. parkeri* in North America. The Asian branch gave rise to the rest of the tick typhus rickettsiae.

The insectborne rickettsiae clearly arose from a tickborne ancestor. The trombiculidborne rickettsiae, however, are so distinct from the other members of the tribe that a close phylogenetic relationship cannot be postulated with great confidence. At best, *R. tsutsugamushi* may be considered to have diverged from a common ancestor of the tickborne group.

2. *Tribe Ehrlichieae*

a. Ehrlichia

The Tribe Ehrlichieae contains organisms pathogenic for certain vertebrates, but (in contrast to the Tribe Rickettsieae) not for man. Species of the type genus, *Ehrlichia*, cause disease in domestic ruminants and dogs. *Ehrlichia canis*, the cause of malignant canine rickettsiosis, is transmitted by the cosmopolitan dog tick, *Rhipicephalus sanguineus*. The other recognized species in *Bergey's Manual*, *E. phagocytophia*, is transmitted to cattle by *Ixodes ricinus* ticks. The diseases caused by these organisms appear to be restricted (or at least indigenous) to the Old World, a strange situation considering the worldwide distribution of *R. sanguineus*, but not a unique one as we shall see.

Several other ehrlichia-like organisms have been described: *E. equi* in California and several agents infecting cattle and sheep in Europe and Africa—*E. bovis*, *E. ovina*, and *Cytoecetes ondiri*. Their association with ticks is tenuous at best, although, in many respects, they resemble *E. canis* and *E. phagocytophila* (see Volume II, Chapter 9).

b. Cowdria

The genus *Cowdria* contains only one species, *C. ruminantium*, the etiologic agent of heartwater fever of goats, sheep, and cattle transmitted by *Amblyomma* ticks. The

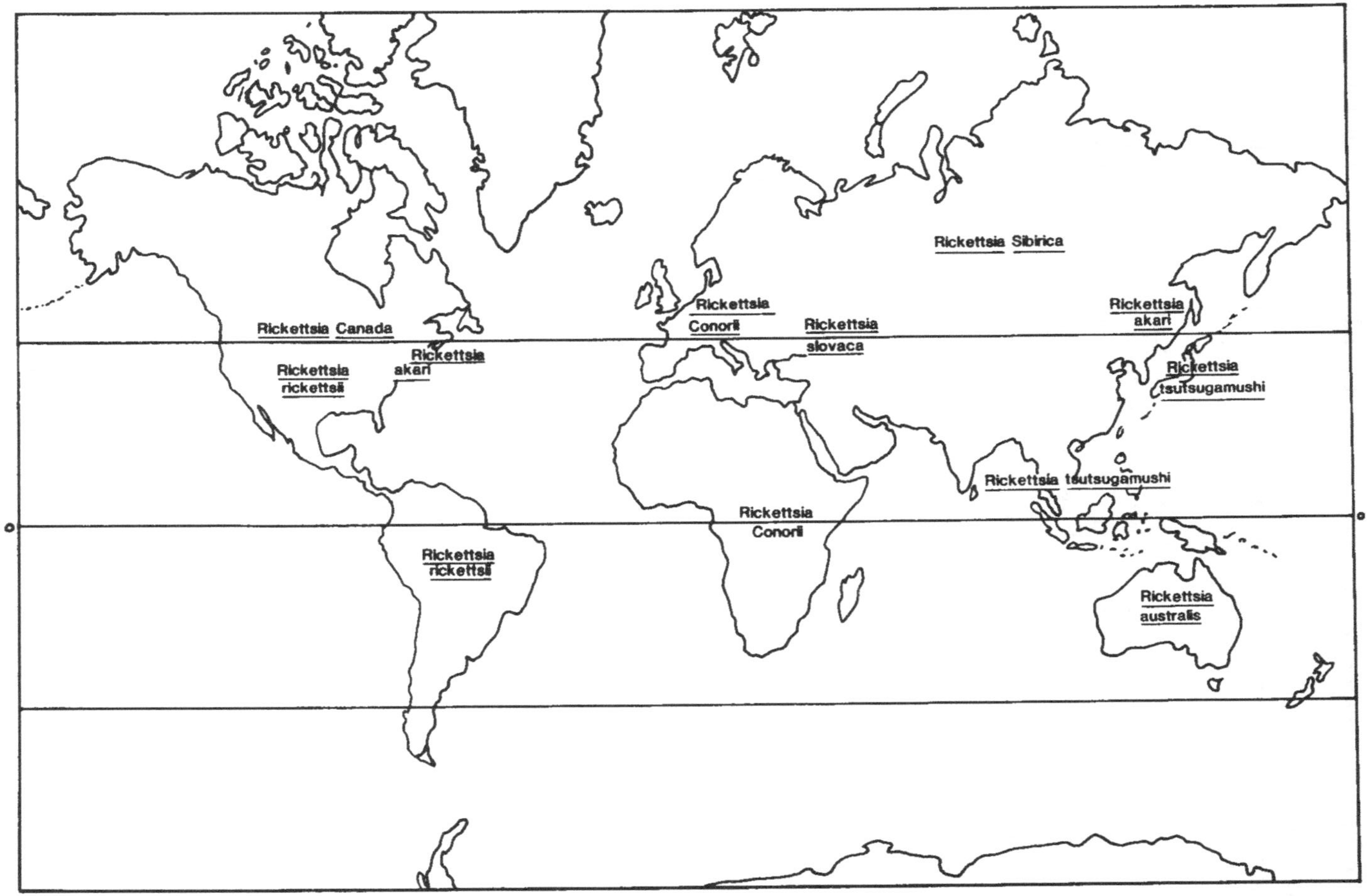

FIGURE 2. Distribution of *Rickettsia*.

organism apparently does not infect the salivary glands of its tick host, and the mechanism of transmission to vertebrates is not known for certain; but contamination of the bite wound with infective tick feces is considered likely. It is passed transstadially, but not transovarially in the tick. It appears to be restricted to continental Africa, but its relationship with *Amblyomma* ticks suggests that it could also exist in South America.

c. Neorickettsia

The third genus in this tribe, *Neorickettsia,* is represented by a single species, *N. helminthoeca,* whose extremely interesting life cycle involves a fluke parasite of salmon and dogs.[48] This organism is harmless to its fluke vector, does not grow at all in the fish host of the immature flukes, and is highly pathogenic to canines, which acquire infection by ingesting salmon with infected flukes. No natural arthropod host is known, but the parasite has been passed experimentally, both transstadially and transovarially, in *Rhipicephalus sanguineus.* It is restricted to the Pacific Northwest corner of the U.S. A detailed comparison of *N. helminthoeca, E. canis, C. ruminantium,* and certain chlamydial agents would, indeed, be interesting.

3. Tribe Wolbachieae

a. Wolbachia

The tribe Wolbachieae, as presently constituted, is a heterogeneous mixture of organisms whose relationships to each other, let alone to the Rickettsieae and Ehrlichieae, are questionable. Steinhaus[49] originally described *Wolbachia dermacentrophila* as a species of *Rickettsia* primarily on its rickettsia-like appearance, its lack of growth on artificial media, and because it was found in *Dermacentor andersoni,* the Rocky Mountain spotted fever tick. In contrast to *W. persica* and the nonpathogenic rickettsia-like symbiotes of hard ticks, *W. dermacentrophila* is not restricted to the Malpighian tubules, but may be found throughout the tick's tissues. Its status as a species distinct from *Rickettsia rickettsii* is in doubt, and it is not a recognized species in *Bergey's Manual.*

Wolbachia persica appears to be a true symbiote of argasid ticks. It is restricted to particular organs (the Malpighian tubules), is found in all stages of its host, is transmitted transovarially (presumably with 100% efficiency), and current evidence suggests that it (or closely related forms) occurs in all argasids.[50–53] Other rickettsia-like symbiotes morphologically similar to *W. persica* also are found in ixodid ticks where they are even more restricted in habitat, being found only in cells of the anterior region of the Malpighian tubules.[53–56]

The other two recognized species of *Wolbachia* are symbiotes of insects, neither of which is pathogenic for vertebrates except possibly *W. melophagi,* the sheep ked organism. Sometimes the association is so close that mycetome like structures are present in the host, but unlike many other insect symbiotes, the microbes are not confined to them.

b. Symbiotes, Blattabacterium **and** ***Rickettsiella***

The rickettsia-like symbiotes of argasid and ixodid ticks resemble, in some respects, *Symbiotes lectularius* in bed bugs (*Cimex* spp.), *Blattabacterium cuenoti* and *Rickettsiella popilliae* in the Japanese beetle (*Popillia japonica,*) but the latter is not a true symbiote since it is pathogenic for the beetle larvae.[57] There is evidence suggesting that *Rickettsiella* may have a complex life cycle similar to that of the Chlamydiaceae,[58] but this work needs confirmation.

Symbiotes lectularius and *Blattabacterium cuenoti* are true symbiotes found in mycetomes in their hosts and also in tissues associated with the digestive tract and reproductive organs. The lack of much basic information on *Symbiotes, Blattabacterium,* and *Rickettsiella* makes it impossible to place them in a phylogenetic classification with the rickettsiae. For present purposes, I am inclined to consider them to have arisen independently of the acarine symbiotes and, until more information is available, to relegate them to the "miscellaneous" category. It is entirely conceivable that *W. persica* and the ixodid Malpighian tubule symbiotes also arose independently of the other rickettsial agents of ixodid ticks, but a phylogenetic relationship between them is equally possible.

B. Bartonellaceae

Weinman[59] has condensed the family Bartonellaceae to two genera, *Bartonella* and *Grahamella*. *Bartonella bacilliformis*, the etiologic agent of oroya fever (*Verruga peruana*), in Peru, Colombia, and Ecuador, is a Gram-negative bacterium cultivable in appropriate media and possesses all the characteristics commonly ascribed to this group of microorganisms. In culture it is motile by means of polar flagellae and differs considerably from *Haemobartonella* and *Eperythrozoon*.[60] Although apparently transmitted by sandflies, there is no evidence that the agent multiplies in the vector.

Grahamella is a poorly studied group of nonpathogenic (or weakly pathogenic) erythrocyte parasites of lower mammals. Their affinities can only be guessed, but the lack of any established association with arthropods would seem to exclude them permanently from the rickettsial evolutionary line.

In my opinion there is little justification for including *Bartonella* and *Grahamella* in the Order Rickettsiales. The only characteristic they share with the rickettsiae is their apparent transmission by arthropods (insects in their case), and *Grahamella* even lacks that. There is no evidence to sugggest that these organisms are even remotely related to the rickettsiae. It might be wise to leave the current genera of the Bartonellaceae in Topley and Wilson's category of "Miscellaneous Bacteria"[61] until sufficient knowledge is acquired to place them in their proper phylogenetic order.

C. Anaplasmataceae

In their revision of the Anaplasmataceae for the Eighth Edition of *Bergey's Manual*, Ristic and Kreier[62] expanded the family to include *Anaplasma, Paranaplasma, Aegyptianella, Haemobartonella*, and *Eperythrozoon*. *Haemobartonella* and *Eperythrozoon* were included in the family Bartonellaceae in the Seventh Edition of *Bergey's Manual*.

1. Anaplasma and *Paranaplasma*

The Anaplasmataceae is a widespread group of parasites of the red blood cells of vertebrates. The genera *Anaplasma* and *Paranaplasma* each contain two recognized species, all transmitted by various species of hard-bodied ticks parasitizing ruminants. They share this latter characteristic with rickettsiae, but differ in being obligate parasites of vertebrate erythrocytes, and may be found in various morphological forms, which according to some workers may represent developmental stages.[63] This latter characteristic is reminiscent of the chlamydiae. On the other hand, parasitization of vertebrate erythrocytes suggested to the early workers a relationship with protozoa.

Species of *Anaplasma* are worldwide in distribution and in some areas are economically important because of the disease they cause in cattle. No known wild vertebrate cycle exists, but the fact that parasitism of bovines is not always benign suggests that cattle may not be the basic vertebrate host.

2. Aegyptianella

Aegyptianella is a group of erythrocyte parasites of birds and the tortoise *Emys leprosa*. The poultry tick *Argas persicas* in all stages is an experimental vector of *A. pullorum*, but it is not known whether other *Aegyptionella* are tick-transmitted.[65–67]

D. Chlamydiales

It is apparent that the widely distributed chlamydiae constitute a group of organisms only remotely related phylogenetically to the Rickettsiales. As Moulder pointed out, the observable similarities between the rickettsiae and the chlamydiae "may be accounted for by the common bacterial lineage of both." They probably evolved along different pathways from a common stem from which most, if not all, the other Gram-negative bacteria also evolved. Both became adapted to intracellular life, but the rickettsiae developed as acarine parasites (or symbiotes) while the chlamydiae differentiated as vertebrate parasites.

A reasonable, if only temporary, solution to the problem of chlamydial taxonomy was proposed by Page[68] who placed all the Chlamydiaceae in two species of the genus *Chlamydia, C. trachomatis,* and *C. psittaci*. This is how they are treated in *Bergey's Manual,* but it is apparent from recent studies that these agents are only remotely related to each other and should be reclassified into separate genera or even separate families. For a more complete discussion of *Chlamydia* nomenclature see Volume II, Chapter 12 in this work.

Much confusion about the true nature of the chlamydial agents stems from the improper and, at times, careless use of terms. Every organism has to have a name, a number, a symbol, or some form of brief descriptive terminology in shorthand. Ideally, the name should convey, at least to the specialist, a clue to the important charcteristics of the organism bearing it. If a newly discovered organism possesses all the important characteristics of a known group it is justifiable to give it the name of that group. The problem is, of course, that during the early phases of group determination, all the significant parameters may not be well defined and the limits of variation may not be established. Thus Lillie[70] originally named the psittacosis agent *Rickettsia psittaci* on the basis of its morphological similarity to known members of the genus *Rickettsia* and its lack of growth on artificial media. Lillie can hardly be criticized for his action since *Rickettsia* was not such a well-defined category of organisms in 1930 as it is today.

In 1931, several inclusion conjunctivitis agents of domestic animals also were given the generic name *Rickettsia.*[71,72] Later, in the 7th edition of *Bergey's Manual,*[1] these and other similar agents were placed in two genera in the family Chlamydiaceae: *Colesiota* Rake, and *Ricolesia* Rake, but Page[69] considers them insufficiently characterized to determined their taxonomic status.

Somewhat later French workers[44,73–75] took a different tack and applied the general name neorickettsia (with a small "n") to a variety of what are probably chlamydial agents in North Africa, but some of which may be *Ehrlichia* or *Rickettsia* in the family Rickettsiaceae.[47,76,77] Unfortunately, *Bergey's Manual* is not universally accepted as the standard work on microbial classification, and the French workers continued to refer to members of the Chlamydiaceae as neorickettsiae, although Giroud[78] acknowledged the inappropriateness of the term.

Finally, the confusion was compounded when Philip and coworkers named a new *Ehrlichia*-like agent of salmon flukes pathogenic for dogs *Neorickettsia helminthoeca.*[79] Japanese workers[80] named a similar agent *N. sennetsu*.

III. DEFINITION OF THE RICKETTSIACEAE

For purposes of establishing a natural system of classification Rickettsiaceae may be defined simply as those Gram-negative bacteria which are usually intracellular, but may be facultatively extracellular, symbiotes, or parasites of arthropods (primarily acarines), usually transmissable to vertebrates, and for which there is reasonably good evidence for phylogenetic relationship. Included under this narrow definition are *Rickettsia, Ehr-*

lichia, and certain *Wolbachia*. Excluded from the family Rickettsiaceae is the Tribe Wolbachieae except for *Wolbachia dermacentrophila*, which may be considered a species of *Rickettsia*. The remaining *Wolbachia* species are insect symbiotes that probably arose independently of the acarine parasites, although perhaps also from soil-inhabiting ancestors.

Ehrlichia canis is the only species of the Tribe Ehrlichieae about which there is adequate information to suggest a phylogenetic relationship to Rickettsieae. Future investigations may show that *E. bovis*, *E. ovina*, and other species deserve inclusion on phylogenetic grounds, but the available evidence is equivocal. I am inclined to exclude *Neorickettsia helminthoeca* unless or until a more definitive relationship to *Rickettsia* or *Ehrlichia canis* can be established.

Two organisms, *Rochalimaea quintana* and *Wolbachia melophagi*, are considered here to have arisen independently from the other groups and probably are related more closely to some other Gram-negative bacterial group, which ultimately, as did all bacteria, originated from primitive soil- or water-inhabiting ancestors.

IV. ORIGIN AND EVOLUTION OF THE RICKETTSIAE

Burnet[81] considered that the "rickettsiae as we know them represent a secondary evolution from a group of ancient, well-adapted parasites or Symbionts"* (referring to the large group of organisms in arthropods, sometimes associated with a special organ, the "mycetome"). He posed the following evolutionary stages:

1. Saprophytic bacteria colonizing the gut of insects
2. Parasitization of the gut epithelium
3. Development of tolerated symbiotes with a latent pathogenicity in the presence of different host cells or conditions
4. Parasitization of vertebrates via blood-sucking insects

While I agree in principle with Burnet's proposal, I differ in considering that rickettsiae and many of the great variety of other arthropod symbiotes have evolved more or less independently of each other. Indeed, symbiotic associations of microorganisms with invertebrates seem likely to have arisen many times. It is only by assuming such a polyphyletic origin that one can explain symbiotic associations involving such diverse microorganisms as fungi, bacteria, rickettsiae, viruses, and protozoa.

The almost complete lack of experimental evidence relating the present pathogenic rickettsiae to the many so-called nonpathogenic rickettsia-like symbiotes of arthropods, some contained in specially developed structures (mycetomes) and most restricted to specific organs, forces us to begin by speculating. The development of the proposed thesis necessitates postulating that the microbial symbiotes of insects and other invertebrates arose quite independently from those of the Acarina, except for those usually acarine symbiotes that can be shown to have adapted secondarily to insects (such as *Rickettsia prowazekii* in lice and *R. typhi* in fleas.)

*Symbiosis refers to the living together of dissimilar organisms and includes all forms of parasitism or commensalism.[82] It is used here in its broadest sense, and I do not attempt to distinguish it from forms of mutualism. In many instances of intracellular symbiosis, we simply do not know whether or not the symbiotes are living in a mutually advantageous association. The terms "symbiote" and symbiont" are used interchangeably to refer to either member of a symbiotic association. As Steinhause[83] pointed out in a more detailed discussion, symbiote is preferable to symbiont since it is etymologically derivable from the original Greek word *symbiotes*, meaning "one who lives with."

For reasons that are not entirely clear but may in part be nutritional, almost all wood-consuming, plant sap-sucking, and blood-feeding arthropods are hosts of symbiotic microorganisms.[84] The adaptation of the ancestral ticks (already containing microbial symbiotes) to blood-feeding on vertebrates probably occurred during the evolution of reptiles in the Mesozoic Era. The decline of the reptiles and rise of mammals forced the parasitic acarines (or provided them with the opportunity) to adapt to new, warm-blooded hosts. Primordial rickettsiae symbiotic in the gut of ixodids then began a stage of adaptive radiation on a worldwide scale, and this adaptation was to both invertebrate and vertebrate hosts:

1. To various local tick-host associations to produce the geographical forms related directly to *R. rickettsii*
2. To the mesostigmatic rodent mites such as *Liponyssoides* producing *R. akari*
3. Through rodent hosts to trombiculid mites, associated with the tremendous development of *Rattus* and the trombiculid genus *Leptotrombidium* in Southeast Asia, with the emergence of *R. tsutsugamushi*
4. Through rodent hosts to blood-sucking insects, accompanied by confinement to the gut, as *R. typhi*, which, by a succession of adaptations, has produced *R. prowazekii*

Secondary extensions to other hosts may be expected and have apparently been obtained in *R. typhi* (through house rats to ticks in India and to trombiculid mites, *Neoscheongastia indica* in Sumatra) and perhaps in *R. prowazekii* (through domestic animals to ticks in Ethiopia).

Baker[85] proposed a similar "epidemic series" for the typhus group of fevers, but his approach was through the development of epidemic disease caused by pathogenic rickettsiae transmitted by arthropods to man and was strongly oriented to the human diseases of typhus, spotted fever, and Q-fever. His sense of evolution was influenced by the nature of the disease in man and he even considered that ". . . the fevers from ticks appear to be relatively young diseases in contrast with a series like the typhus series." My approach to the evolution of tickborne rickettsiae is more biological, more from the point of view of the rickettsia, if you will. The diseases in man caused by these organisms, while of undeniably great public health importance, are quite irrelevant to the evolution of the rickettsiae with the possible exception of *Rickettsia prowazekii*. Even in this instance, however, some evidence indicates that man may be less important to the maintenance of the organism in nature than was previously thought. The evidence for (and against) this view will be developed in some detail in Volume I, Chapter 6.

The arthropods involved in rickettsial evolution are two different groups of insects and of acarines: (1) lice (*R. prowazekii*) and fleas (*R. typhi*); (2) parasitiform ticks or mites (*R. rickettsii, R. akari*), and trombiculid mites (*R. tsutsugamushi*). This peculiar host distribution of the rickettsiae happens to be associated with an ability (or inability) to penetrate the gut-wall of the arthropod vector and often to pervade its tissues generally.

With the general outline of rickettsial evolution in mind we can now examine the evidence on which the hypotheses are based. In the following chapters, I hope to be able to show how the pathogenic rickettsiae could have evolved in the course of geologic time through a series of evolutionary developments involving acarines and vertebrates as well as the microbes themselves. The study of parallel evolution of host and parasite as a tool in clarifying the evolution of the parasite is not a unique approach.

Metcalf[86] used it effectively in the study of opalinid ciliate infusoria,* Kellogg[87] in studying the evolution of ectoparasites, and Darling[88] in the study of hookworms.

REFERENCES

1. **Philip, C. B.,** Class III. Microtatobiotes, in *Bergey's Manual of Determinative Bacteriology,* 7th ed., Breed, R. S., Murray, E. G. D., and Smith, N. R., Eds., Williams & Wilkins, Baltimore, 1957, 931.
2. **Philip, C. B.,** Comments on the classification of the Order Rickettsiales, *Can. J. Microbiol.,* 2, 261, 1956.
3. **Allison, A. C. and Burke, D. C.,** The nucleic acid contents of viruses, *J. Gen. Microbiol.,* 17, 181, 1962.
4. **Schaechter, M., Tousimis, A. J., Cohn, Z. A., Rosen, H., Campbell, J., and Hahn, F. E.,** Morphological, chemical and serological studies of the cell walls of *Rickettsia mooseri, J. Bacteriol.,* 74, 822, 1957.
5. **Wissig, S. L., Caro, L. G., Jackson, E. G., and Smadel, J. E.,** Electron microscopic observations on intracellular rickettsiae, *Am. J. Pathol.,* 32, 1117, 1956.
6. **Stelzner, A. and Linss, W.,** Binary fission in *Coxiella burneti, Nature (London),* 218, 1069, 1968.
7. **Bovarnick, M. R. and Schneider, L.,** The incorporation of glycine-1-C^{14} by typhus rickettsiae, *J. Biol. Chem.,* 235, 1727, 1960.
8. **Buchanan, R. E. and Gibbons, N. E.,** Eds., *Bergey's Manual of Determinative Bacteriology,* 8th ed., Williams & Wilkins, Baltimore, 1974, 1.
9. **Levaditi, J. C., Roger, R., and Destombes, P.,** Tentative de classification des Chlamydiaceae (Rake 1955) tenant compte de leurs affinities tissulaires et de leur epidemiologie, *Ann. Inst. Pasteur,* 107, 656, 1964.
10. **Krieg, A.,** *Grundlagen der Insektenpathologie. Viren-, Rickettsien- und Bakterien-Infektionen,* Steinkopff, Darmstadt, 1961, 1.
11. **Moulder, J. W.,** *The Psittacosis Group as Bacteria,* John Wiley & Sons, New York, 1964, 1.
12. **Moulder, J. W.,** The relation of the psittacosis group (chlamydiae) to bacteria and viruses, *Annu. Rev. Microbiol.,* 20, 107, 1966.
13. **Moulder, J. W.,** The rickettsias, in *Bergey's Manual of Determinative Bacteriology,* 8th ed., Buchanan, R. E. and Gibbons, N. E., Eds., Williams & Wilkins, Baltimore, 1974, 882.
14. **Weiss, E.,** Growth and physiology of rickettsiae, *Bacteriol. Rev.,* 37, 259, 1973.
15. **Palmer, E. L., Mallovia, L. P., Tzianabos, T., and Obijeski, J. F.,** Electron microscopy of the cell wall of *Rickettsia prowazeki, J. Bacteriol.,* 118, 1158, 1974.
16. **Plotz, H., Smadel, J. E., Anderson, T. F., and Chambers, L. A.,** Morphological structure of rickettsiae, *J. Exp. Med.,* 77, 355, 1943.
17. **Weiss, L. J.,** Electron micrographs of rickettsiae of typhus fever, *J. Immunol.,* 47, 353, 1943.
18. **Mudd, S. and Anderson, T. F.,** Pathogenic bacateria, rickettsias and viruses as shown by the electron microscope, *JAMA,* 126, 561, 632, 1944.
19. **Shkolnik, I. F., Zatulovsky, B. G., and Shestopalova, N. M.,** Ultrastructure of *Rickettsia prowazeki.* An electron microscope study of ultrathin sections from infected louse guts and chick embryo yolk sacs, *Acta Virol. Eng. Ed.,* 10, 260, 1966.
20. **Glauert, A. M., Brieger, E. M., and Allen, J. M.,** The fine structure of vegetative cells of *Bacillus subtilis, Exp. Cell Res.,* 22, 73, 1961.
21. **Bayer, M. E. and Anderson, T. F.,** The surface structure of *Escherchia coli, Proc. Natl. Acad. Sci. U.S.A.,* 54, 1592, 1965.
22. **Petris, S. de,** Ultrastructure of the cell wall of *Escherchia coli* and the chemical nature of its constituent layers, *J. Ultrastruct. Res.,* 19, 45, 1967.
23. **Anderson, D. R., Hopps, H. E., Barile, M. F., and Bernheim, B. C.,** Comparison of the ultrastructure of several rickettsiae, ornithosis virus and mycoplasma in tissue culture, *J. Bacteriol.,* 90, 1387, 1965.

*Opalinid ciliate infusoria are protozoan symbiotes in the gut of frogs and toads. They live in the rudimentary coecal portion of the rectum and occur in almost every frog and toad.

24. **Anacker, R. L., Pickens, E. G., and Lackman, D. B.,** Details of the ultrastructure of *Rickettsia prowazeki* grown in the chick yolk sac. *J. Bacteriol.*, 94, 260, 1967.
25. **Popov, V. L. and Ignatovich, V. F.,** Electron microscopy of surface structure of *Rickettsia prowazeki* stained with ruthenium red, *Acta Virol. Engl. Ed.*, 20, 424, 1976.
26. **Higashi, N.,** Recent advances in electron microscope studies on ultrastructure of rickettsiae, *Zentralbl. Bakteriol. Parasitenkd. Infektionskr. Hyg. Abt. 1 Orig.*, 206, 277, 1968.
27. **Silverman, D. J. and Wisseman, C. L., Jr.,** Comparative ultrastructural study on the cell envelopes of *Rickettsia prowazekii, Rickettsia rickettsii* and *Rickettsia tsutsugamushi, Infect. Immun.*, 21, 1020, 1978.
28. **Silverman, D. J., Wisseman, C. L., Jr., Waddell, A. D., and Jones, M.,** External layers of *Rickettsia prowazekii* and *Rickettsia rickettsii:* occurrence of a slime layer, *Infect. Immun.*, 22, 233, 1978.
29. **Schaechter, M., Bozeman, F. M., and Smadel, J. E.,** Study on the growth of rickettsiae. II. Morphologic observations of living rickettsiae in tissue culture cells, *Virology*, 3, 160, 1957.
30. **Salton, M. R. J.,** Studies of the bacterial cell walls. IV. Composition of some Gram-positive and Gram-negative bacteria, *Biochim. Biophys. Acta*, 10, 512, 1953.
31. **Allison, A. C. and Perkins, H. R.,** The presence of cell walls like those of bacteria in rickettsiae, *Nature (London)*, 188, 796, 1960.
32. **Perkins, H. R. and Allison, A. C.,** Cell wall constituents of rickettsiae and psittacosis-lymphogranuloma organisms, *J. Gen. Microbiol.*, 30, 469, 1963.
33. **Wisseman, C. L., Jr.,** Some biological properties of rickettsiae pathogenic for man, *Zentralbl. Bakteriol. Parasitenkd. Infektionsk. Hyg. Abt. 1 Orig.*, 206, 299, 1968.
34. **Manire, G. P. and Tamura, A.,** Preparation and chemical composition of the cell walls of mature infectious dense forms of meningo-pneumonitis organisms, *J. Bacteriol.*, 94, 1178, 1967.
35. **Costerton, J. W., Ingram, J. M., Cheng, K. J.,** Structure and function of the cell envelope of gram negative bacteria, *Bacteriol. Rev.*, 38, 87, 1974.
36. **Dhir, S. P., Hakomori, S., Kenny, G. E., and Grayston, J. T.,** Immunochemical studies on chlamydial group antigens, *J. Immunol.*, 109, 116, 1972.
37. **Lewis, V. J., Thacker, W. L., and Witchell, S. H.,** Demonstration of chlamydial endotoxin-like activity, *J. Gen. Microbiol.*, 114, 215, 1979.
38. **Costerton, J. W., Poffenroth, L., Wilt, J. C., and Kordova, N.,** Ultrastructural studies of *Chlamydia psittaci* 6BC "in situ" in yolk sac explants and L cells: a comparison with gram negative bacteria, *Can. J. Microbiol.*, 21, 1433, 1975.
39. **Matsumoto, A. and Manire, G. P.,** Electron microscopic observations of the fine structure of cell walls of *Chlamydia psittaci, J. Bacteriol.*, 104, 1332, 1970.
40. **Garrett, A. J., Harrison, M. J., and Manire, G. P.,** A search for the bacterial mucopeptide component, muramic acid, in *Chlamydia, J. Gen. Microbiol.*, 80, 315, 1974.
41. **Kordova, N.,** Chlamydiae, rickettsiae and their cell wall defective variants, *Can. J. Microbiol.*, 24, 339, 1978.
42. **Pinkerton, H.,** Pathogenic rickettsiae with particular references to their nature, biologic properties and classification, *Bacteriol. Rev.*, 42, 37, 1942.
43. **Cuenod, A. and Nataf, R.,** Bacteriological and experimental researches on the etiology of trachoma, *Br. J. Ophthalmol.*, 21, 309, 1937.
44. **Giroud, P. and Jadin, J.,** Premiers resultats concernant le virus des Bashi isole dans la province du Kivu au Congo Belgie, *Bull. Soc. Pathol. Exot.*, 47, 578, 1954.
45. **Eddie, B., Meyer, K. F., Lambrecht, F. L., and Furman, D. P.,** Isolation of ornithosis bedsoniae from mites collected in turkey quarters and from chicken lice, *J. Infect. Dis.*, 110, 231, 1962.
46. **Meyer, K. F.,** The host spectrum of psittacosis-lymphogranuloma vererum (PL) agents, *Am. J. Ophthalmol.*, 63, 1225, 1967.
47. **Giroud, P. and Colas-Belcour, J.,** Infection neo-rickettsiennes de l'homme apres piqure de *Dermacentor marginatus, Bull. Soc. Pathol. Exot.*, 50, 194, 1957.
48. **Philip, C. B.,** There's always something new under the "parasitological" sun (the unique story of helminth-borne salmon poisoning disease), *J. Parasitol.*, 41, 125, 1955.
49. **Steinhaus, E. A.,** Rickettsia-like organism from normal *Dermacentor andersoni* Stiles, *Public Health Rep.*, 57, 1375, 1942.
50. **Suitor, E. C. and Weiss, E.,** Isolation of a rickettsia-like microorganism (*Wolbachia persica* n. sp.) from *Argas persicus* (Oken), *J. Infect. Dis.*, 108, 95, 1961.
51. **Roshidy, M. A.,** Observations by electron microscopy and other methods on the intracellular rickettsia-like microorganism of *Argas persicus* Oken (Ixodoidea, Argasidae), *J. Insect. Pathol.*, 3, 148, 1961.
52. **Roshidy, M. A.,** Rickettsia-like microorganisms in the Malpighian tubules and ovary of *Argas boeuti* Roubaud and Colas-Belcour, *A. vespertilionis* Latreille, *A. transgariepinus* White and *A. brumpti, Proc. Egypt. Acad. Sci.*, 17, 74, 1964.

53. **Roshidy, M. A.,** Rickettsia-like symbiotes in ticks, *Acarologia,* 6, 274, 1964.
54. **Mudrow, E.,** Uber die Intracellularen Symbionten der Zecken, *Z. Parasitenkd.,* 5, 138, 1932.
55. **Till, W. M.,** A contribution to the anatomy and histology of the brown ear tick, *Rhipicephalous appendiculatus* Neumann, *Men. Entomol. Soc. South Afr.,* 6, 1, 1961.
56. **Cowdry, E. V.,** A group of microorganisms transmitted hereditarily in ticks apparently unassociated with disease, *J. Exp. Med.,* 41, 817, 1925.
57. **Dutky, S. R. and Gooden, E. L.,** *Coxiella popilliae* n. sp., a rickettsia causing blue disease of Japanese larvae, *J. Bacteriol.,* 63, 743, 1952.
58. **Huger, A. M. and Krieg, A.,** New aspects of the mode of reproduction of *Rickettsiella* organisms in insects, *J. Invertebr. Pathol.,* 9, 442, 1967.
59. **Weinman, D.,** The rickettsias, Family II. Bartonellaceae Gieszczykiewicz, 1939, in, *Bergey's Manual of Determinative Bacteriology,* 8th ed., Buchanan, R. E. and Gibbons, N. E., Eds., Willimas & Wilkins, Baltimore, 1974, 903.
60. **Peters, D. and Wigand, R.,** Bartonellaceae, *Bacteriol. Rev.,* 19, 150, 1955.
61. **Wilson, G. S. and Miles, A. A.,** *Topley and Wilson's Principles of Bacteriology and Immunology,* 5th ed., Edward Arnold, London, 1964, 1093.
62. **Ristic, M. and Kreier, J. P.,** The rickettsias, Family III. Anaplasmataceae (Philips 1957), in *Bergey's Manual of Determinative Bacteriology,* 8th ed., Buchanan, R. E. and Gibbons, N. E., Eds., Williams & Wilkins, Baltimore, 1974, 906.
63. **deRobertis, E. and Epstein, B.,** Electron microscope study of anaplasmosis in bovine red blood cells, *Proc. Soc. Exp. Biol. Med.,* 77, 254, 1951.
64. **Tanaka, H., Hall, W. T., Sheffield, J. B., and Moor, D. H.,** Fine structure of *Haemobartonella muris* as compared with *Eperythrozoon coccoides* and *Mycoplasma pulmonis, J. Bacteriol.,* 90, 1735, 1965.
65. **Bedford, H. A. H. and Coles, J. D. W. A.,** The transmission of *Aegyptianella pullorum* Carpano to fowls by means of ticks belonging to the genus *Argas, Onderstepoort J. Vet. Sci. (Anim. Ind.),* 1, 15, 1933.
66. **Gothe, R.,** Ein Beitrag zur systematischen Stellung von *Aegyptianella pullorum* Carpano, 1928, *Z. Parasitenkd.,* 29, 119, 1967.
67. **Hadani, A. and Dinur, Y.,** Studies on the transmission of *Aegyptianella pullorum* by the tick *Argas persicus, J. Protozool.,* 15 (Suppl.), 45, 1968.
68. **Page, L. A.,** Proposal for the recognition of two species in the genus *Chlamydia* Jones, Rake and Stearns, 1945, *Int. J. Syst. Bacteriol.,* 18, 51, 1968.
69. **Page, L. A.,** Revision of the Family Chlamydiaceae Rake (Rickettsiales): unification of the psittacosis-lymphogranuloma venereum-trachoma group of organisms in the genus *Chlamydia* Jones, Rake and Stearns, 1945, *Int. J. Syst. Bacteriol.,* 16, 223, 1966.
70. **Lillie, R. D.,** Psittacosis: rickettsia-like inclusions in man and in experimental animals, *Public Health Rep.,* 45, 773, 1930.
71. **Coles, J. D. W. A.,** A rickettsia-like organism in the conjunctiva of sheep, *17th Rep. Dir. Vet. Serv. Onderstepoort,* Sec. II, 1931, 175.
72. **Coles, J. D. W. A.,** A rickettsia-like organism of the conjunctival epithelium of cattle, *J. S. Afr. Vet. Med. Assoc.,* 7, 221, 1936.
73. **Giroud, P., Roger, F., and Dumas, N.,** Contribution a l'etude des neorickettsioses. L'evolution des anticrops au cors des diverses maladies de l'homme et des animaux, *Bull. Soc. Pathol. Exot.,* 48, 21, 1955.
74. **Jadin, J. and Giroud, P.,** Constations epidemiologiques et serologiques sur les neo-rickettsies, *Acad. R. Sci. Colon. (Brussels), Sect. Sci. Natl. Med. Nouvelles Serie,* 7, 1, 1957.
75. **Jadin, J., Leonard, J., and Thomas J.,** Neo-rickettsies et assortement chex les bovids en Belgique, *C. R. Soc. Biol.,* 154, 1127, 1960.
76. **Rousselot, R.,** *Rickettsia (Donatienella) delpyi* n. sp., n. subgen., *Bull. Soc. Pathol. Exot.,* 41, 110, 1948.
77. **Giroud, P., Boyer, J., and Vargues, R.,** Les ricketsioses latent. Deductions epidemiologiques et prophylactiques, *Paris Med.,* 40, 413, 1950.
78. **Giroud, P.,** Des agents de la psittacose a ceaux du trachome, *Presse Med.,* 77, 475, 1969.
79. **Philip, C. B., Hadlow, W. J., and Hughes, L. E.,** *Neorickettsia helmintheca,* a new rickettsia-like disease agent of dogs in western United States transmitted by a helminth, *Abst. 6th Int. Microbiol. Congr.,* Vol. 2, Int. Congr. Microbiol., Rome, 1953, 256.
80. **Misao, T. Y. and Kobayashi, Y.,** Studies on infectious mononucleosis (glandular fever). I. Isolation of etiologic agent from blood, bone marrow, and lymph node of a patient with mononucleosis by using mice, *Kyushu J. Med. Sci.,* 6, 145, 1955.
81. **Burnet, F. M.,** The rickettsial diseases of Australia, *Med. J. Aust.,* 2, 129, 1942.

82. **Hertig, M., Taliaferro, W. H., and Schwartz, B.,** The terms symbiosis, symbiont and symbiote, *J. Parasitol.*, 23, 326, 1937.
83. **Steinhause, E. A.,** *Insect Microbiology*, Comstock, Ithaca, N.Y., 1946, 1.
84. **Koch, A.,** Intracellular symbiosis in insects, *Annu. Rev. Microbiol.*, 14, 121, 1960.
85. **Baker, A. C.,** The typical epidemic series, *Am. J. Trop. Med.*, 23, 559, 1943.
86. **Metcalf, M. M.,** The opalinid ciliate infusorians, *U.S. Natl. Mus. Bull.*, 120, 1, 1923.
87. **Kellogg, V.,** Distribution and species-forming of ectoparasites, *Am. Nat.*, 47, 129, 1913.
88. **Darling, S. T.,** Observations on the geographical and ethnological distribution of hookworms, *Parasitology*, 12, 217, 1920.

Chapter 3

EVOLUTION OF THE TICK-RICKETTSIA RELATIONSHIP

N. Marchette

TABLE OF CONTENTS

I. Acarina 30
  A. Origin of the Acarina 30
  B. Divergence of Argasidae and Ixodidae 33

II. Microbial Symbiotes of Ticks 33

III. Origin and Development of Tick-Rickettsia Relationship 36

IV. Phylogeny and Distribution of Ticks 37
  A. Argasidae 38
  B. Ixodidae 38
    1. *Aponomma* 38
    2. *Amblyomma* 39
    3. *Margaropus* 39
    4. *Boophilus* 39
    5. *Hyalomma* 40
    6. *Rhipicephalus* 40
    7. *Dermacentor* 41
    8. *Haemaphysalis* 41
    9. *Ixodes* 42

V. Summary 43

References 45

I. ACARINA

The Order Acarina has been popularly subdivided into ticks and mites on the basis of size, and elementary textbooks tend to perpetuate this idea separating the ticks too strictly from the closely related parasitoid mites. The Acarina in fact may be divided into two large assemblages according to the refractility of their chitin[1] and on the structure of their digestive, excretory, and respiratory systems. Thus, the anactinochitinose (nonrefractile chitin) acarines comprise the ticks and what are often called parasitoid or gamasoid mites (Mesostigmata), including the common "laelaptid" rodent mites *Liponyssoides sanguineus* (natural vector of *Rickettsia akari*) and *Ornithonyssus bacoti* (experimental vector of *R. akari*). They are often incorporated into the suborder Parasitiformes on morphological grounds, expecially the position of the stigmata and the structure of the gnathosoma. The ticks or Metastigmata (Table 1) are a wholly parasitic and specialized group that probably evolved from a common stem with the parasitoid mites. The latter include (1) many predatory species, (2) a number of species and higher taxa parasitic on vertebrates, and (3) some species parasitic on other arthropods. Arthropod parasitism by other arthropods is usually considered to be derived secondarily.[6]

In constrast, the actinochitinose (refractile chitin) acarines are a larger and much more heterogeneous group including sarcoptiform and trombidiform mites, which may, however, be more closely related to each other than are the gamasoid mites and ticks. The trombiculid vectors of scrub typhus belong to this group.

Also included in the Acarina are the Onychopalpida, a group of primitive mites unique in that they have claws on the pedipalps. They are of no known significance to rickettsial ecology or evolution, perhaps because nothing is known of their microbial symbiotes.

A. Origin of the Acarina

The age of the planet Earth is in excess of five billion years, but life as we know it probably originated less than a billion years ago in the vast oceans that covered much of the globe. Colonization of the land began approximately half a billion years ago in the Paleozoic Era. The evolution and diversification of vertebrates has taken place mostly during the last 250 million years or so in the Mesozoic and Cenozoic Eras. The major geologic time divisions are shown in Table 2. The Acarina probably arose in the Devonian.[7] Recent discovery of fossil terrestrial arachnids in the Rhynie Chert deposits in Germany suggests that nonscorpion arachnids were well established on the land in lower Devonian time and developed contemporaneously with land plants and insects.[8] All the higher orders appeared then and during the Carboniferous period. They apparently differentiated mostly under terrestrial conditions, which were adverse to their preservation,[9] and this accounts, in part, for the sparse fossil record of the terrestrial groups.

The Metastigmata (ticks) and the Mesostigmata (containing the gamasid mites) are postulated to have arisen from a common stem, the parasitic members tracing their ancestry back to soil-inhabiting forms. The majority of the Mesostigmata are still free-living in soil and decaying organic matter. The Metastigmata diverged from the common stem with the Mesostigmata by adapting completely to a parasitic existence on vertebrates. The appearance of terrestrial vertebrates thus signaled the evolution of ticks. The Mesozoic era in which they arose was the Age of the Reptiles,[10] the first wholly terrestrial vertebrates. Presumably ticks achieved a high degree of development and diversity, perhaps almost equalling that of their reptilian hosts, during this immensely long period.[11] Unfortunately, we have no direct evidence for this because ticks are not readily preserved and do not appear in the fossil record of the Mesozoic. Fossil

Table 1
CLASSIFICATION OF THE IXODOIDEA[2–5]

Class Arachnida
Subclass Acari
Superorder Anactinochaeta
Order Metastigmata

Suborder Ixodidea							
Family Ixodidae					Family Argasidae		Family Nuttalliellidae
Ixodes Latreille	*Margaropus* Karsch *Boophilus* Curtice *Rhipicentor* Nuttall and Warburton *Nosomma* Schulz *Cosmiomma* Schulze *Dermacentor* Koch *Rhipicephalus* Koch *Anomalohimalaya* Hoogstraal, Kaiser, and Mitchell	*Hyalomma* Koch	*Amblyomma* Koch *Aponomma* Neumann	*Haemaphysalis* Koch	*Argus* Latreille	*Ornithodoros* Koch *Antricola* Cooley and Kohls *Otobius* Banks	*Nuttalliella* Bedford

Note: The tick genera within each column are more closely related to each other than they are to those in the other columns.

Table 2
MAJOR GEOLOGICAL TIME DIVISIONS

Era	Period	Epoch	Approx. no. of years before present (millions)
Cenozoic	Quarternary	Recent	0.025
		Pleistocene	2
	Tertiary	Pliocene	13
		Miocene	25
		Oligocene	35
		Eocene	58
		Paleocene	63
Mesozoic	Cretaceous		135
	Jurassic		181
	Triassic		230
Paleozoic	Permian		280
	Carboniferous		345
	Devonian		405
	Silurian		425
	Ordovician		500
	Cambrian		600

ticks from any period are extremely rare. Weidner[12] described a species of *Ixodes* very similar to *I. ricinus* embedded in Baltic amber, fossilized resin from extinct coniferous trees that flourished along the Baltic coast in the Eocene. Many amber specimens contain free-living mites, but this is the only one in which a tick has been found. An *Ixodes* tick has been described from remains found in the Green River, Wyoming Oligocene beds, but there is some question that it is even a tick. The only other fossil tick is a male *Dermacentor* found in the exterior ear canal of a wooly rhinoceros in Pliocene deposits in Poland.[12]

A well-developed terrestrial vertebrate fauna was probably necessary for the evolution of ticks from their free-living ancestors. Primitive amphibians were the first terrestrial vertebrates and their divergence from strictly aquatic forms in the Devonian[10] may have coincided roughly with a similar divergence of terrestrial arachnids. However, the early land arachnids belonged to the orders Acarida, Anthracomasti, and Araneida; the Acarina probably developed somewhat later, diverging from already well-established arachnids. Present-day amphibians generally possess few acarine parasites, and their early ancestors presumably were not involved in the evolution of ticks or mites. The appearance of a fauna of wholly terrestrial vertebrates would seem to have been necessary for the development of parasitic ticks. Reptiles were the first vertebrates capable of completing their entire life cycle on land, and they almost certainly played an essential role in tick evolution.

At the very least, toward the end of the Age of Reptiles, a well-developed and widely distributed fauna of reptilian tick parasites must have existed. With the decline of the reptiles at the end of the Mesozoic, many of the more highly specialized ticks probably followed their hosts to extinction. The only modern tick genus that is essentially confined to reptiles is *Aponomma,* but it is not clear if the association is an ancient one of primary evolutionary significance, or if it is secondarily derived, perhaps as an offshoot of the *Amblyomma* line. The vast majority of *Aponomma* ticks are reptile parasites, but a few species also parasitize other animals. Australian *Aponomma* occur normally on the spiny anteater and on the wombat as well as on various species of reptiles. A few species have been collected from domestic animals on rare occasions.[13]

During the decline of the reptiles or even before, the more adaptable or more generalized ticks adapted to the hairy or feathery homeothermous vertebrates that first appeared during the Permian and Triassic or Jurassic periods and were destined, after nearly 100 million years, to become the dominant forms of vertebrate life on earth. Very primitive and tiny mammal-like creatures are thought to have arisen as early as the Permian[14] and may be the ancestors of the insectivores. Birds evolved later in the Triassic or Jurasic period. These major evolutionary events took place after the development of reptiles from amphibian ancestors during the Carboniferous, but long before the reptiles reached their peak diversity and distribution. Thus, both birds and mammals evolved from reptilian stock while reptiles were still in the dynamic phases of development, and these diminutive creatures apparently represented a minor portion of the terrestrial vertebrate fauna until the reptiles began their decline at the end of the Mesozoic. Conceivably, acarines developed initially as parasites of primitive mammals (or birds) and diverged later to parasitize reptiles. This point may never be answered satisfactorily, but at the time acarines first appeared, reptiles, in terms of numbers and diversity of species alone, might have been a better biological system for the evolution of blood-feeding ectoparasites.

The important point is that some, or all, of the current tick genera probably arose by adaptation of reptilian ticks to mammals and birds during the Mesozoic. The dawn of the Cenozoic Era witnessed the increasingly rapid development of the mammalian fauna which was destined to replace the reptiles as the dominant terrestrail vertebrates. The major development, divergence, and dispersal of mammals throughout the world occurred during the Eocene. This process was probably most spectacular among the Rodentia, which were the dominant Eocene mammals in terms of number, variety, and distribution.

B. Divergence of Argasidae and Ixodidae

The obligate parasitic existence of ticks makes it axiomatic that they were deeply involved with the development of the warm-blooded vertebrates. It is not clear precisely when the Argasidae (soft-bodied ticks) and Ixodidae (hard-bodied ticks) or their progenitors diverged, but it probably occurred very early. A plausible time would have been during the latter part of the Mesozoic, possibly as early as the Jurassic when birds are thought to have evolved, some tens of millions of years after the appearance of the first primitive mammals in the Permian. A line of ticks became adapted to parasitic existence on birds and nidocolous mammals and eventually evolved into the present-day family Argasidae. The other main line of ticks, and the one with which we are most concerned, were more adaptable. They parasitized a great variety of mammals and evolved into the present-day family Ixodidae.[11]

II. MICROBIAL SYMBIOTES OF TICKS

If we assume that primitive reptile ticks already possessed rickettsia-like symbiotes before the two major families diverged, then those in the Argasidae for some reason failed to "develop" further and remained strict symbiotes in their tick hosts (for example, *Wolbachia persica* in *Argas arboreus*). They are confined to the Malpighian tubules and developing oocytes and do not penetrate the gut wall; thus they cannot reach the salivary glands or coxal fluid to be transmitted to the tick's vertebrate hosts,[15] and they apparently are not shed in the feces. Roshidy[16] considers all the rickettsia-like symbiotes of argasid ticks to be related to the one studied by him in *Argas arboreus*.*

The uniformity and universality of *W. persica* and "related" symbiotes in the Argasidae may indicate an early initial association between rickettsia and tick. Steinhause[18] called attention to the possible relation of symbiotes to the systematic position of their arthropod hosts and cited examples from Buchner[19] showing that in some insect groups every member has the same type of symbiote, even up through the suborders. In other groups the unifomity of symbiote types extends only to the family level, in still others only to subfamily. The implication is that taxa with uniform symbiote types probably acquired their symbiotes before differentiating from the parent stem. Thus we might say the Argasidae acquired the *Wolbachia*-like symbiotes before or during their differentiation from the Ixodidae.

If the rickettsial symbiotes of the Ixodidae (including those in the Tribe Rickettsieae) are related to the Wolbachieae of the Argasidae, the symbiotic relationship between ticks and rickettsia-like agents would have occurred before the differentiation and separation of Argasidae and Ixodidae, which we have placed in the Jurassic or early Cretaceous, coinciding with the appearance of birds. The similarity of the food habits of the argasids and ixodids (all are obligate blood feeders), however, cannot be overlooked as possibly inducing a parallel development of symbiotes in the two families. Such specialization of microbes to obligate blood-feeding arthropods might also account for the apparent similarity between the various rickettsia-like organisms that occur in such diverse insects as mosquitoes, fleas, and lice, all of which are now placed in the genus *Wolbachia*.[20] Thus, relating these insect symbiotes to the *Wolbachia* of ticks may not be phylogenetically sound.

The argasids are, of course, important vectors of spirochaetes,[21] some of which cause relapsing fever in man and other vertebrates. In some ticks these bacteria are transmitted transovarially and their presence may have excluded rickettsiae transmissible to vertebrates from that particular niche through competition early in their evolutionary history. This line of argument is purely conjectural and several apparent facts seem to contradict it. Many argasids continue to parasitize birds, but little evidence suggests the occurrence of widespread tickborne spirochaetal infections in wild birds although spirochaetoses do occur. Also, few spirochaetes are known from reptiles. Thus, one might argue that the argasid-spirochaete relationship developed in conjunction with argasid tick parasitization of mammals, suggesting that the argasid-spirochaete association is somewhat more recent than the ixodid-rickettsial relationship. This may well be the case since *Borrelia* species occur mainly in *Ornithodoros* ticks, which are primarily mammal parasites.** An exception is a fowl spirochaetosis transmitted by *A. persicus*.

***Wolbachia persica* was first described in *Argas (Persicargas) arboreus* not in *A. (P.) persicus* as its name implies. There has been a great deal of confusion between these two species, but the situation has been clarified.[17]

**Epidemic relapsing fever of man is louseborne and caused by *Borrelia recurrentis;* endemic relapsing fever is tickborne and caused by various other *Borrelia* species. The louseborne and tickborne relapsing fever spirochaetes are essentially indistinguishable morphologically. It is generally conceded that *B. recurrentis* is derived from *B. duttoni*, the *Borrelia*-louse cycle perhaps arising as an accident of nature following louse feeding on a tick spirochaete infected host. *Borrelia duttoni* is carried by *Ornithodoros moubata*, a tropical East African tick that has become highly domiciliated in and around African huts. All the tickborne relapsing fever spirochaetes are naturally transmitted by species of *Ornithodoros* and *Argas* ticks; other *Borrelia* are transmitted by boophilids and perhaps other ixodids. The New World *Borrelia* appear to be host specific, and each species is generally found in and is transmitted only by its natural host. The Old World *Borrelia*, particularly *B. duttoni* and *B. hispanica*, are less host specific and are capable of developing in ticks other than their natural vectors and also in human lice.[22–25]

The transovarial transmission of *Borrelia* by argasid ticks apparently does not inhibit the transovarial transmission of *Wolbachia* or vice versa in these same ticks. Also, species of hard ticks (*Dermacentor, Haemaphsyalis, Ixodes, Hyalomma, Rhipicephalus,* and *Boophilus*) which transmit rickettsiae and anaplasmas to vertebrates, transmit piroplasms, too. Sometimes the same species (such as, *Rhipicephalus sanguineus*) is an important vector of both organisms. Riek[26] and Shortt[27] have demonstrated hereditary transmission of the protozoons in *Boophilus, Dermacentor, Haemaphysalis, Hyalomma, Rhipicephalus,* and *Ixodes* ticks.*

Why one family of ticks (Argasidae) apparently does not carry rickettsiae transmissable to vertebrates and another family (Ixodidae) does has not been satisfactorily explained. This may not be entirely true if *Aegyptianella* (in the family Anaplasmataceae) is considered a rickettsia. *Argas* ticks have been shown to transmit it.[30,31] One can postulate rather lamely that the early adaptation of the Argasidae to birds and nest- or burrow-inhabiting mammals had something to do with it. Or, to put it the other way around, the adaptation of the Ixodidae to free ranging mammals favored the development of a gut wall penetrable by rickettsiae, or favored the development of rickettsiae that could penetrate the gut wall and infect the salivary glands. Rehacek[15] discusses some potential factors involved, but comes to no definite conclusions. The so-called "gut-wall barrier" that confines *Rickettsia* spp. to the gut lumen and intestinal epithelium of argasids appears to be real, but no one has yet explained the mechanism by which it operates.

Burgdorfer[32] showed that *Coxiella burnetii* can survive in *Ornithodoros moubata* for more than a year, but this tick is incapable of biological transmission of the organism. *Coxiella burnetii* has been isolated from the coxal fluid of parenterally inoculated *O. moubata*[33] and *Rickettsia australis* from the hemolymph of parenterally infected ticks of the same species.[34] Also, Kordova and Rehacek[35] found *R. prowazekii* in various organs of parenterally infected ticks, indicating the ability of at least some rickettsiae to grow and multiply in extra-gut tissues of ticks that do not normally transmit them. *Cowdria ruminantium* propagates in epithelial gut cells of *Amblyomma hebraeum,*[36] apparently without passing through the gut wall.** Since this tick is apparently the sole vector of heartwater fever of goats, sheep, and cattle in Africa, some other mechanism of transmission must exist, perhaps fecal contamination of the bite wound. The other rickettsial diseases of ruminants and other domestic animals (caused by species of *Ehrlichia*) in Africa and Europe are thought to be transmitted by various ixodid ticks.[37]

*The Theileriidae are generally believed not to be transovarially transmitted in the vector ticks.[28] The Babesiidae, on the other hand, all pass through the egg except for *Nuttallia tadzhikistanica,* which is passed only transstadially.[29] Riek[26] lists the following ticks in which *Babesia* is transovarially transmitted: *Boophilus annulatus, B. decoloratus, B. microplus, Rhipicephalus bursa, R. sanguineus, Ixodes ricinus, I. persulcatus, Dermacentor marginatus, D. silvarum, D. nitens, Hyalomma dromedarii, H. marginatum, H. excavatum, Haemaphysalis leachi,* and *H. bispinosa.*

**Hurlbut and Thomas[38] provided further evidence for the existence of a gut wall impervious to microorganisms and succeeded in demonstrating the ability of parenterally infected *Ornithodoros savignyi* to support the growth of viruses for which they are not natural vectors. In Storey's classical work on the bushystunt virus of maize transmitted by the leaf-hopper (*Cicadulina mbila* Naude), he reported that certain hoppers were unable to transmit the virus after ingestion, although they became effective vectors when the gut wall was perforated with a fine needle.[39] The inability to transmit the virus was due to blockage of the passage of the virus through the gut wall, and this was found to be a dominant sex-linked hereditary characteristic. The work of Rochow[40] suggests that the vector specificity of barley yellow dwarf virus is a function of the virus capsid, and it is now known that this is true for many animal viruses. Philip[41] presents other situations of this kind, and, to add to the complexity, Smith[42] demonstrated that *Coxiella burnetii* cannot pass through the gut wall of *Haemaphysalis humerosa* (a hard-bodied tick) in spite of propagation in gut epithelial cells; it can, however, cross the gut wall of many other ixodid ticks.

Much work has been done on various aspects of the relationship of organisms to the gut wall of arthropods, and a critical review would be valuable. Among the factors known to be involved in specific cases are the presence of a peritrophic membrane (absent in ticks) and of specific antibiotic agents apparently produced by the epithelial cells. An important point for investigation is the fate of organisms when their ingestion is shortly followed by metamorphosis with its accompanying changes. All rickettsial agents so far as known are transstadially transmitted.

One important aspect of the acarine-rickettsia relationship must be borne in mind. The only facts available are those discernible from the fauna now living, which is the current product of more than 150 million years of evolution. Conceivably the avian fauna at some time in its history was actually infected with tickborne rickettsiae much as mammals are today, then later became refractory to infection by the argasid symbiotes. Or it may be that we have simply missed rickettsial parasites of birds. As long as the rickettsiae can infect succeeding tick generations by transovarian transmission, a vertebrate cycle may not be essential. Whatever the reason, we know of no true rickettsial pathogens of birds transmitted exclusively (or even primarily) by argasid ticks.

III. ORIGIN AND DEVELOPMENT OF TICK-RICKETTSIA RELATIONSHIP

Either some rickettsial symbiotes of the Ixodidae clearly did develop the ability to penetrate the gut wall, infect the salivary glands, and eventually multiply in vertebrate hosts, or the gut wall of ixodid ticks developed in such a way that it became permeable to some gut-dwelling rickettsial symbiotes. It is also possible that changes in both symbiotes and ticks occurred during the process of adaptation and development so that the symbiotes were no longer confined to the gut epithelium. These are all ways of saying that something happened, but we have no idea of the mechanism involved or how it came about. Microorganisms that infect vertebrates and appear in their blood conceivably would have a selective advantage over those not so adapted. Additional generations of rickettsiae would be produced, possibly increasing the chance that a favorable mutation would occur. The wide geographical range of the vertebrate host would serve to increase the distribution of the rickettsiae and make it available to other vertebrate parasites, which might become suitable hosts of the rickettsia and thus increase not only its distribtuion but also its host range.

The alternative hypothesis—that the ancestors of present-day rickettsiae capable of infecting vertebrates were initially picked up by ixodid ticks from their naturally infected vertebrate hosts—must also be entertained. The discovery of rickettsia-like organisms in tick-free guinea pig colonies[43] and the isolation of rickettsia-like agents from a water rat in Australia[44] may suggest a vertebrate origin for the rickettsiae, which became adapted to ticks only after the evolution of mammals. However, these agents are probably unrelated to the rickettsiae (see Volume II, Chapter 7). Other workers have reported the isolation of *Chlamydia* from laboratory colonies of guinea pigs[45–47] and mice.[48] Since the relationship of tickborne rickettsiae to the so-called rickettsia-like agents or chlamydiae of guinea pigs and rats is unlikely, the acarine origin of rickettsiae remains the most logical hypothesis.

Despite the existence of possible exceptions to the original hypothesis, ultimate sources of the parasitic rickettsiae undoubtedly were free-living bacteria in the soil, a vast bacteriological laboratory sometimes containing several tons of bacterial substance per acre. Also, a tremendous wealth of arthropods exist in the soil and have repeated opportunities to ingest soil organisms either directly because they feed on them, or indirectly because they prey on or parasitize herbivorous arthropods. Most species are herbivorous, parasitic, or predatory; very few are truly saprophytic.[49]

A chain of adaptations may therefore be postulated whereby free-living soil bacteria adapted to soil arthropods, became facultative parasites, and, ultimately, obligatory parasites. Ticks apparently do not have a chance to acquire organisms from such a source for they are consistently parasitic on vertebrates. Nevertheless, ticks evolved from the mesostigmatic stem, a large assemblage of mites that are primarily predators but have often and in many groups independently adopted parasitism on both arthropods and vertebrates. Therefore, if the tickborne rickettsiae constitute the primitive stem of *Rickettsia,* which is the present contention, then it must have arisen when ticks evolved from the common mesostigmatic stem. We might therefore expect to find representatives among the mesostigmatic mites. Unfortunately, there has been relatively little effort along this line, and it is not at all clear that the gamasid mite-transmitted *R. akari* represents a primitive organism.

On the other hand, trombiculid mites appear to have had a much better chance of acquiring rickettsial precursors because they feed mainly on other soil arthropods. Also, although their parasitic larvae now feed on vertebrates, the trombiculids have evolved from a common stem with the other trombidiform mites, among which the parasitic larvae attack arthropods and so may also ingest a variety or organisms. This provides some ground for speculating that the trombiculidborne rickettsiae may be the primitive rickettsial stem. This hypothesis would be strengthened if more rickettsiae in trombiculids were found in Africa and the Americas. The possibility of parallel evolution of *R. tsutsugamushi* and the tickborne *Rickettsia* spp. cannot be entirely ruled out and will be examined in Volume II, Chapter 7.

The rickettsiae in ticks and parasitoid mites comprise those responsible for the various forms of tick-typhus and rickettsialpox in man. A peculiarity of some members of this group is that they may invade and multiply in the nucleus of various host cells,[50] but the implications of this are not known. From their present relationships to each other and to vertebrates and arthropods, we can follow the evolutionary development of *Rickettsia* from the most primitive tickborne *Rickettsia* species to the most highly specialized insectborne *Rickettsia*. This is harder to do with the other so-called "rickettsiae", such as *Coxiella burnetii, Anaplasma,* and *Cowdria*. They probably are not rickettsiae at all in the phylogenetic sense. This will be discussed in later chapters.

Whatever the ultimate origin of the Rickettsiales, there is little disagreement that the development and dispersion of ticks have had a profound effect on the evolution and speciation of at least one major branch of the Tribe Rickettsieae.

IV. PHYLOGENY AND DISTRIBUTION OF TICKS

The Metastigmata is a group of obligate ectoparasites that feed on blood and tissue fluids of vertebrates. Ticks are worldwide in distribution but have reached their greatest diversity and development in the Old World Tropics—the Ethiopean and Oriental Faunal Regions of Sclater,[51] Wallace,[52] and Schmidt[53] (see Figure 1, Chapter 2.) They are easily separated into two distinct families, the Ixodidae and the Argasidae* (see Table 1), on the basis of the rigidity of their cuticle and the presence or absence of a dorsal scutum. The Ixodidae or hard-bodied ticks have a comparatively rigid cuticle and possess a sclerotized dorsal shield or scutum. They are slow feeders, have only one nymphal instar, and their developmental cycle may require one, two, or three ver-

*A third family, Nuttalliellidae, containing a single species of the genus *Nuttalliella* is of little importance to rickettsial evolution.

tebrate hosts depending upon the species. All species of *Boophilus* are one-host ticks, many species of *Rhipicephalus* require two to three hosts, and most others are three-host ticks, each instar requiring a different host.[11]

A. Argasidae

The Argasidae or soft-bodied ticks lack a dorsal scutum and have a leathery cuticle. Their developmental cycle may include as many as eight nymphal instars. They may be considered multihost ticks, the rapid-feeding nymphs and adults taking a number of short blood meals and spending most of their life in nests of their hosts. Although there is some difference of opinion, four genera of argasid ticks are generally recognized.[54] *Argas* and *Ornithodoros* are distributed throughout the world. Most species of *Argas* parasitize birds and bats, but some also feed on lizards, ground mammals (including *A. echinops* which parasitizes insectivores), and one is a parasite of the Galapagos giant tortoise.[55] *Ornithodoros* ticks are parasites of mammals as well as birds and reptiles. These two genera contain most of the more than 160 recognized species of Argasidae. There are two species of the genus *Otobius*, which feed on a variety of domestic and wild mammals. They are found mostly in the New World (Neotropical and Nearctic Faunal Regions), but have also been reported on domestic animals in South Africa and India. The five species of the small genus *Antricola* are known only from bats in the southwestern U.S., Mexico, Central America, and the Caribbean. Argasid ticks are primary arthropod hosts of relapsing fever spirochaetes and are known to be hosts of many arboviruses, but they are only incidentally involved in the ecology of rickettsiae and rickettsia-like organisms with the exception of *Aegyptianella pullorum*.

B. Ixodidae

Ixodid ticks also are important transmitting hosts of arboviruses, but they are of greatest interest here as the basic maintaining and transmitting hosts of rickettsiae and in the role they have played in the evolution of the Rickettsiaceae. The small and geographically restricted *Cosmiomma* and *Nosomma* of Schulze[56] are closely related to *Hyalomma* and will not be distinguished from it here. *Indocentor* (Schulze) and *Rhipicentor* (Nuttall and Warburton), included by some in the genus *Dermacentor*, are important only to tick taxonomy, which will be discussed only very superficially. Interested readers may refer to the publications of Nuttall and Warburton[57] and Sharif.[3]

1. Aponomma

Species of *Aponomma* parasitize reptiles almost exclusively. This might be fundamentally important from the viewpoint of the early evolution of tickborne rickettsiae with the possibility that they were transmitted to reptiles much as they now are to mammals. Unfortunately, little is known about rickettsiae in *Aponomma* and almost nothing about rickettsiae in reptiles. The few tantalizing bits of information available are of uncertain significance. For instance, there are reports of experimental infection of turtles with *Coxiella burnetii*,[58,59] and Kazar et al.[60] have reported that *R. akari* and *R. conorii* grow to higher titer in primary tissue culture of colubriid liver cells than in chick embryo yolk sacs. The infected snake cells do not appear to be damaged in any way by the rickettsiae and reproduce normally. In contrast Schindarow[61] reported that *R. conorii* multiplies slowly in cultures of tortoise kidney cells and produces slight cytophathic effect. In this experiment, compact extranuclear colonies formed in the infected cells. *Rickettsia prowazekii* and *R. typhi* do not grow well in these cell cultures.

Species of *Aponomma* are more numerous in the Oriental Faunal Region than in Africa. Only four species exist in the Americas and none in Europe.[62] This might be taken as further evidence of the evolution of ticks in the Old World Tropics, but there is no proof that *Aponomma* is a more primitive genus than the other ixodids. The genus *Amblyomma,* to which *Aponomma* is closely related, also contains many species parasitic on reptiles as do other genera, such as the immature stages of a few *Haemaphysalis* species.

2. *Amblyomma*

Amblyomma is widely distributed throughout the Old and New World approximately between the 40th parallels of latitude.[63] About half the established species are found in the Americas, mostly in South America, with Africa next in order of species abundance. Asia and Australia have only a few species each. Immature amblyommids are common on birds and small mammals; the adults generally parasitize herbivores and rarely carnivores. In Asia, wild pigs are important hosts. As noted above, some species also parasitize reptiles (for example, *A. helvolum* and *A. goeomydae*) and such scaled mammals as the scaly anteater or pangolin, *Manis javanicus,* of Southeast Asia (*A. javanense*). Some South American species have very specialized host preferences.

One might speculate that the present distribution of *Amblyomma,* mainly in tropical South America and Africa, is a contraction of a once world-wide distribution on reptiles before the appearance of rodents and other mammals as the dominant vertebrates. They may have failed to adapt as readily as the other ixodid lines to the new, more metabolically active creatures and continued to maintain their dominance in South America where only one group of rodents (the Hystricomorpha) are thought to have evolved, along with a rich fauna of marsupials. The present fauna of placental mammals, with the exception of the hystricomorphs, are of North American origin. Their persistence in Africa may be explained by a late adaptation to the larger herbivores and carnivores which have evolved to such a high degree on that continent. The *Aponomma* line persisted on the rich reptilian fauna of tropical Asia.

The alternate and preferable hypothesis is that *Amblyomma* evolved in the Southern Hemisphere (Gondwanaland supercontinent) and was never represented in the Northern Hemisphere any more than it is now. It may also be significant that *Amblyomma* is represented in Australia by more species than any other genus except *Ixodes*.[64,65] We might also speculate that the involvement of amblyommids with pathogenic rickettsiae is secondarily derived. The evidence regarding these views will be presented later.

3. *Margaropus*

The small genus *Margaropus* is confined to Africa and, as far as can be determined, has not been implicated in transmission of rickettsial pathogens of vertebrates. Members of the genus contain symbiotes in their Malpighian tubules, as do all ixodids, but nothing is known of their relationships, if any, to the genus *Rickettsia*.

4. *Boophilus*

The genus *Boophilus* with its five species is almost as small as *Margaropus,* but it is world-wide in distribution. It is related to *Rhipicephalus* and *Anocenter* and probably originated as a parasite of ungulates.[66] The five known species, *B. annulatus, B. decoloratus, B. microplus, B. kohlsi,* and *B. geigyi,* are one-host ticks parasitizing deer, antelope, and domestic herbivores. They transmit various species of Piroplasms;[67] a virus has been isolated from *B. microplus* collected from cattle in Singapore and Malaysia;[68] and they are experimental and probably natural hosts of *Anaplasma*. Gear[69] mentions that *B. decoloratus* ". . . has been stated to be a vector (of *R. conorii*) and may well be, but no strain has been isolated from it." Except for *B. kohlsi,* which parasitizes sheep and goats in arid Jordan,[70] the boophilids are generally restricted to humid tropical and subtropical regions. Earlier reports of *B. decoloratus* in the Near

East were actually *B. kohlsi*. *Boophilus annulatus* is a North American tick and vector of the protozoan, *Babesia bigemina,* etiological agent of Texas fever. *Boophilus microplus* is widely distributed in the tropics and subtropics of southern Florida, Central America, South America, Australia, the Orient, and East Africa. *Boophilus decoloratus* and *B. geigyi* are African species.

5. *Hyalomma*

The original center of dispersion of *Hyalomma* was probably the Middle East in or near what is now Iran or southern Russia.[62] The *Hyalomma* ticks of Russia, the Middle East, and North Africa appear to be rather closely related, but they differ widely from a special group in Pakistan and India.[71] This might suggest a long period of isolation of these two fauna due to the barriers posed by the Himalayan mountains to the north of India and the Khyber range between West Pakistan and Afganistan. However, many common species occur both North and South of the Himalayas. The evolutionary history of *Hyalomma* is obviously much more complex than it appears at first glance.

6. *Rhipicephalus*

Africa is the place of origin and center of distribution of *Rhipicephalus*. According to Hoogstraal[62] in his monumental work on the ticks of the Sudan, 39 species are endemic in the Ethiopean Faunal Region. This region includes all of Africa south of the southern edge of the Sahara Desert.* Only two are Oriental, and five range from southern Europe and North Africa into Russia.

The dog tick *R. sanguineus* has been distributed around the world by man. *Rhipicephalus sanguineus, R. simus, R. appendiculatus* and *R. evertsi* are the prime vectors of *Rickettsia conorii* (at least in cycles involving man) in Africa and the Mediterranean littoral. In India *R. sanguineus* is believed to transmit Indian tick typhus, but the usual vector may well be *R. turanicus*. These species are often confused, and *R. sanguineus*** is very localized in India, whereas *R. turanicus* is much more common.[72] In the warmer parts of the Americas *R. sanguineus* occasionally may be a vector of Rocky Mountain spotted fever. It also transmits *Ehrlichia canis* and *Babesia canis* to dogs.

The many rhipicephalid species (60 or more) in continental Africa and the taxonomic problems they present led Kaiser and Hoogstraal[73] to surmise that they may be ". . . a comparatively recent, rapidly evolving group." The rich and varied large vertebrate fauna of Africa may account for the great diversity of *Rhipicephalus* species there. An almost equally large and varied herbivore population once existed in South and Southeast Asia, but the rhipicephalids apparently were not there to develop as they did in Africa.

*Some authors place the Sahara in the Palearctic region, some call it an intermediate faunal region.

**The complex *Rhipicephalus sanguineus* group has been studied by Morel and Vassiliades[74] and separated into five distinct species, only two of which are known to transmit rickettsiae (*R. sanguineus* and *R. turanicus*). Most are parasites of carnivores in the adult stage. All stages of *R. sanguineus* are found on dogs, but immature stages may also parasitize rodents, and adults often parasitize domestic herbivores as well as carnivores. It has been distributed throughout the world on domestic dogs. *Rhipicephalus pusillus* is a specific parasite of the warren rabbit, *Oryctolagus cuniculus,* and is confined to the western Mediterranean region. It occasionally parasitizes rodents and carnivores in the vicinity of rabbit warrens. The natural distribution of *R. guilhoni* is in the north Sudanese savannahs and the Sahelian steppe from the Atlantic to eastern Chad; that of *R. sulcatus* in the southern Sudanese and Guinean savannahs. *Rhipicephalus turanicus* is found from the Mediterranean region to China where the adults parasitize herbivores and carnivores: the immature stages feed on rodents.

7. Dermacentor

Species of *Dermacentor* ticks transmit a number of viral and rickettsial diseases to man. They are common in the temperate zone and are poorly represented in the tropics. Only three species are found in tropical Africa,[62] and the genus is represented in tropical Asia by the *D. auratus* group comprised of at least three distinct species:[75] *Dermacentor atrosignatus* Neumann the most common, *D. compactus* Neumann next most common, and *D. auratus* Supino, rarely found in tropical Asia east of India. All parasitize wild pigs in the adult stage.

The center of origin and distribution of *Dermacentor* is probably central or north central Asia where they transmit not only members of the Russian Spring Summer encephalitis group of viruses, but are also the prime vectors of *Rickettsia sibirica*. In western North America, *D. andersoni* is an important vector of *R. rickettsii*, but *D. variabilis* and possibly *D. parumapertus* also play a role as do other ixodids to a lesser extent.

8. Haemaphysalis

The haemaphysalids are to Asia, where more than 100 out of a total of approximately 155 species occur, what the rhipicephalids are to Africa. Eight of the twenty species of ixodids known from the Philippines are haemaphysalids.[76–78] In South India, 14 species of *Haemaphysalis* are known from the Kyasanus Forest region in Shimoga district alone;[79] 8 of them parasitize monkeys. Approximately 16 species exist in tropical and South Africa, but the yellow dog tick, *H. leachii*, is by far the most common.[62] Hoogstraal[62] has stated that there are about five species each in the Americas and Europe, but Hohls[80] lists only three species in the New World, and about the same number are found in Europe; there are seven species in Australia.[81]

In the Madagascan Archipelago, nine of the ten endemic ticks are haemaphysalids (the other one is *Ixodes lunatus* Neumann), most of them related to Oriental forms.[82] This is in harmony with geological and other evidence indicating the Asian rather than African affinity of Madagascar.*

Until the last two decades, the *Haemaphysalis* ticks of the Palearctic region have been much more intensively studied than those of the Oriental region, yet in the whole of the U.S.S.R., Pomerantzev[86] lists only 11 species. Teng[87] reported 24 species in China, but half of them are distributed in the south and southwest.

Hoogstraal[88] postulated that *Haemaphysalis* ticks evolved as reptile parasites in the late Paleozoic or early Mesozoic in tropical Southeast Asia. Structurally primitive forms are relatively rare, and in only a few species do the immature stages feed on lizards. The adults almost never parasitize reptiles. Until 1966, only one such primitive form, *H. theilerae* of Madagascar, was known from the Tropics; all the others are confined to temperate zones of North America and Eurasia.[89] Then Hoogstraal and Wilson[90] discovered and described what they considered as ". . . possibly the most structurally primitive of all contemporary haemaphysalids" from the highlands of Vietnam.

Any discussion of *Haemaphysalis* ticks must remain incomplete until the publication

*Conceivably, Madagascar was separated from Africa during the early stages of breakup of Gondwanaland. It may have remained attached to the Indian subcontinent for a long period before breaking loose as India drifted north. For an even longer period, numerous islands may have formed a complex chain of stepping stones between Madagascar and India. The Seychelles, Amirantes, Agalegas, Farquhars, Tromelin, La Reunion, Mauritius, and others may be present-day remnants of such an island chain. The unique fauna of Madagascar is difficult to explain on any other basis. However, a dissenting view is presented by Millot[83] and Fooden,[84] and the reconstruction of Dietz and Holden[85] shows Madagascar still attached to Africa long after India began its northward migration (see Volume I, Chapter 4).

Table 3
DISTRIBUTION OF MAJOR SUBGENERA OF HAEMAPHYSALIS

Subgenus	Approx. no. of species	India[a]	South-east Asia[b]	Mada-gascar	Pale-arctic	Nearctic	Ethi-opian	Aust-ralia
Aboimisalis	4	1			1	2		
Aborphysalis	5	2	4					
Alloceraea	5	1	1		3			
Allophysalis	5	3			4			
Haemophysalis	25	13	12		3			
Herpetobia	5	5						
Kaiseriana	32	14	15		5		1	2
Ornithophysalis	17	4	5	2	1		2	6
Rhipistoma	22	4	6	7	2		6	
Dermaphysalis	1			1				
Sharifiella	1			1				
Gonixodes	2					2		
Total	124	47	43	11	19	4	9	8

[a]All of India and adjoining regions from the Himalayas south and including Assam.
[b]Including Taiwan, Indonesia, and the Philippines.

of Hoogstraal's epic revision of the genus. Although admittedly incomplete, Table 3 may give some idea of the complexity of the genus and its distribution. These data were taken from the various publications of Hoogstraal and colleagues.[65,75,77–79,81,90–143]

Haemaphysalids are vectors of *R. conorii* in Africa, *R. rickettsii* in the Americas, *R. sibirica* in Russia, and they are presumed to carry tick typhus rickettsiae in Malaysia.[144] *Haemaphysalis* ticks also are major vectors of Kyasanur Forest disease virus in India.

9. *Ixodes*

There is a consensus among tick taxonomists that the genus *Ixodes* stands more or less by itself as a unique, highly specialized group of the Ixodidae[145,146] (Table 1). If the absence of eyes is a primitive character, as considered by some, then *Ixodes* and *Haemaphysalis* represent the most ancient genera. Neither possesses eyes, but they differ in many other important characteristics. Each stands more or less by itself, showing little relationship to the other genera; they are perhaps more closely related to each other than to any other genus. Whatever the affinities of *Ixodes* to other Ixodidae, it clearly seems to be a highly specialized branch of the Proixodoidae stock and to have diverged very early from *Haemaphysalis* and the other genera.

The center of origin and distribution of *Ixodes* is obscure, but more than 60 species out of about 200 are known in Africa[147] and more than 40 in North America,[148] most of them in the U.S.[149] Anastos[150] lists 24 species from the U.S.S.R. This list of *Ixodes* species is probably not complete, but fewer species appear to exist elsewhere than in Africa and North America. Most African species are found south of the Sahara; only 12 have been found in all of North Africa and the Near East as far as the Iraq-Iran border. Relatively few *Ixodes* are known in South America, but this was also true of Africa only a short time ago. Monographic treatment of the *Ixodes* in South America might result in a significant increase in the number of species known to occur there. In Australia there are more species of *Ixodes* (21)* than any other genus.[65,151]

Ixodes ticks have been implicated in transmission of *Babesia bovis,* but it is in the field of virology that this genus assumes its greatest public health importance. Louping ill, a virus-induced poliomyelitis disease of sheep in Great Britain and Europe, has long been known to be transmitted by *Ixodes ricinus.*[153] This tick is also the vector of central European tickborne encephalitis of central and northern Europe and biphasic meningoencephalitis of Russia. Russian Spring Summer encephalitis (RSSE) virus in Russia is transmitted by *I. persulcatus*. Langat virus, a member of the RSSE complex of tickborne viruses, has been isolated only from *I. granulatis* in one locality in West Malaysia[154] and from Thailand.[155]

Few species of *Ixodes* have been involved as vectors of rickettsiae. *Coxiella burnetii* has been isolated from various *Ixodes* species, but this ubiquitous microorganism has also been isolated from *Haemaphysalis, Dermacentor, Amblyomma, Boophilus,* and *Hyalomma* ticks as well as from other arthropods and dozens of mammals and birds. The unique place this organism occupies in rickettsial phylogeny will be discussed in some detail in a later chapter.

Ixodes ricinus and *I. hexagonus* apparently play a role in the ecology of *R. conorii* is France,[156] and *I. ricinus* transmits tickborne fever of cattle and sheep in Europe.

Marchette[144] reported presumptive isolation of tick typhus rickettsiae from each of five pools of adult *Ixodes granulatus* collected from wild rodents in Malaya, but was not able to culture the agent in chick embryos or maintain it in guinea pigs or hamsters. The presence of tick typhus rickettsiae in the tick pools was presumed on the basis of complement-fixing antibody conversion in guinea pigs inoculated with tick suspensions.

Man-biting *Ixodes holocylcus* in Queensland is probably the vector of *Rickettsia australis.*[157,158] It apparently is confined to Australia, and all published records of its occurrence elsewhere are based on misidentifications according to Roberts.[151]

The etiology of the disease known as tickborne paroxysmal rickettsiosis of Ukrainia has not been well characterized, but is reportedly transmitted by *Ixodes ricinus.*[159]

V. SUMMARY

This necessarily brief and incomplete discussion is merely intended to outline the present distribution of genera of the Ixodidae, to speculate on the centers of origin and distribution of each, and to indicate in a general way their relationship to rickettsiae.

Figure 1 shows the current major centers of distribution of ixodid genera. Two things stand out sharply: (1) the richness of the tick fauna in the Old World compared to that in the New World and (2) the concentration of ixodids in the Old World Tropics compared to the rest of the world. *Dermacentor* is clearly a North Temperate Zone genus with major centers of development in Eurasia and North America. *Ixodes* is well represented in North America, but Africa also has a rich assemblage of species. South

*Two primitive subgenera of *Ixodes* (*Sternalixodes* and *Endopalpiger*) are confined to the Australian region where they are exclusively parasites of marsupials. Another primitive group, *Exopalpiger*, is distributed in the Australian, Ethiopian, and European regions. Outside Australia it mainly parasitizes shrews and, less often, mice, rats, and moles. Theiler[152] suggests that shrews, the most primitive of placental mammals, may have been the original hosts of the *Exopalpiger* group outside Australia. In Australia where placental mammals failed to evolve, these three primitive *Ixodes* groups retain their early ancestral habits of parasitizing marsupials. There is no intention to imply that placental mammals arose from marsupials, only that when Australia separated from Gondwanaland, it either did not contain representatives of the placental mammal ancestral stock, or, if it did, they did not survive.

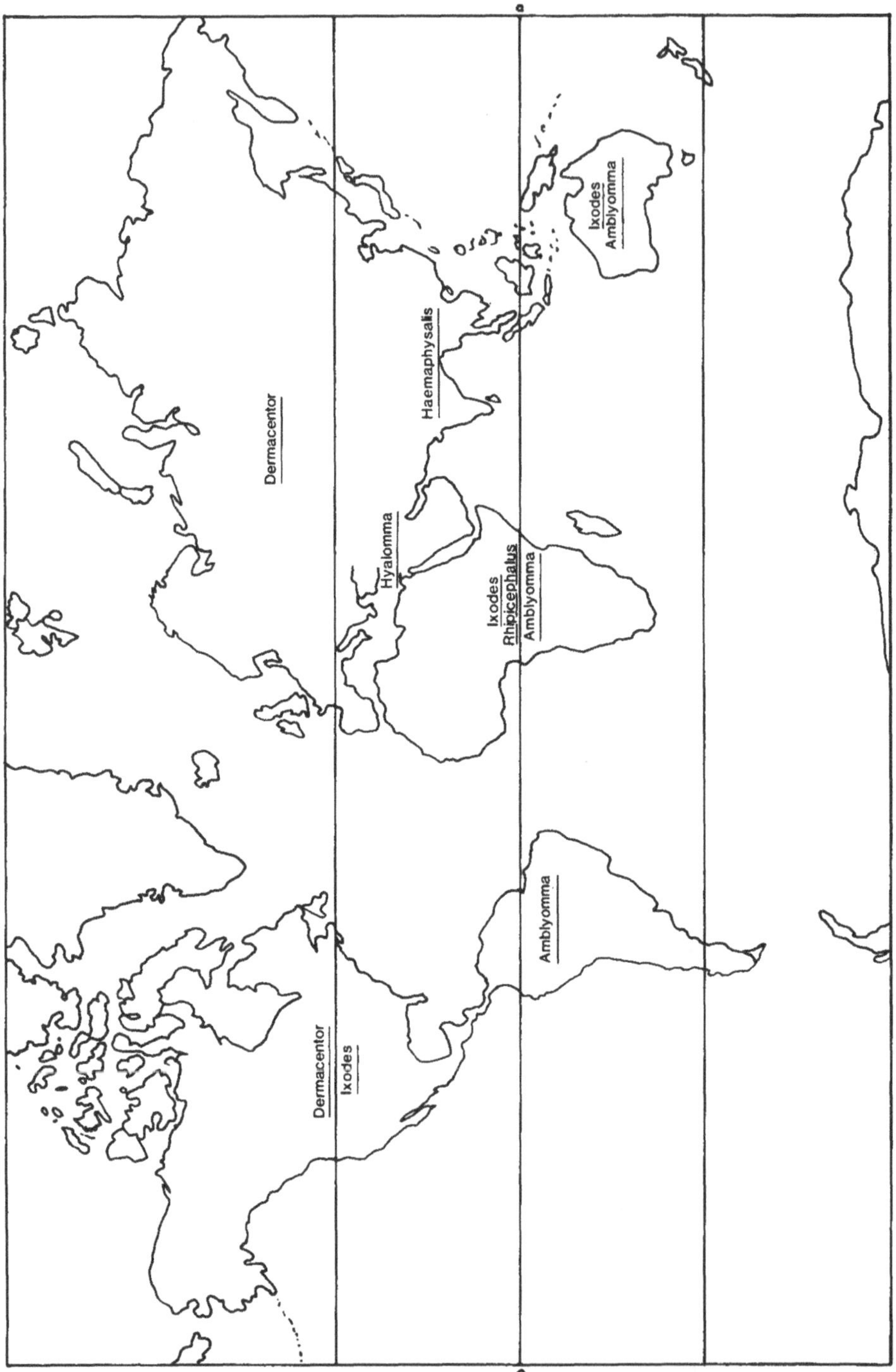

FIGURE 1. Distribution of tick families.

America is now a major center of development of *Amblyomma*, but again Africa has an almost equally rich fauna. *Rhipicephalus* is clearly of African origin; *Hyalomma* is especially well developed in the Middle East; and *Haemaphysalis* lays claim to Southeast Asia as its probable home.

All ixodid ticks possess microbial symbiotes in their Malpighian tubules, but these are true symbiotes hereditarily transmitted from one generation to another and do not infect vertebrates. The rickettsial symbiotes that are transmitted to vertebrates by ticks are not found with equal frequency in all ixodid genera, not even with equal frequency in all species of the genera that possess them. Thus we can narrow down the list of ticks of more than incidental importance in rickettsial evolution to include only six of the ten recognized genera. Three of these, *Haemaphysalis, Dermacentor*, and *Rhipicephalus*, are of major importance; *Hyalomma* and possibly *Amblyomma* may have played a significant but perhaps lesser part. The role of *Ixodes* is somewhat less well established. It is interesting to note that *Dermacentor* and *Rhipicephalus* are considered to have evolved more recently (with the evolution of mammals) than the older genera: *Hyalomma, Amblyomma*, and *Ixodes*.[11]

A single factor does not operate to the exclusion of all others in any biological system. Biological phenomena are vastly complex and dynamic interactions of multitudinous processes. It should be obvious, however, that the present geographical distribution of the tickborne species of rickettsiae can be attributed to the distribution of their tick hosts which, in turn, is determined to a large extent by the distribution of their vertebrate hosts. Rodents are major hosts of most ixodid ticks (at least in their immature stages) involved in rickettsial ecology, but many rodent ticks also spend part of their life cycles (usually the adult stage) on the larger herbivores and carnivores. These creatures with their relatively large home ranges have influenced the distribution of ticks throughout the world and continue to carry individual ticks far from their hatching grounds. The role of birds in the dissemination of ticks must also be seriously considered. Hoogstraal at al.[160,161] showed that migrating birds regularly carry many species of ticks back and forth between Europe, Asia, and Africa, although in another study Nuorteva and Hoogstraal[162] found few ticks on birds migrating to Finland from southern Europe and Asia and none on birds migrating from Africa. Presumably, therefore, migratory birds in other parts of the world also transport ticks over great distances. Relatively few tick species, however, are invloved and not all of them regularly spend part of their life cycle on rodents.

The evolution and development of larger mammals has undoubtedly contributed to the present distribution of ticks that parasitize rodents in one or more stages of their life cycles. However, I am inclined to believe that rodents were the major vertebrates responsible for the distribution of ticks involved in rickettsial evolution. Thus, at this point we are logically led to consider the evolution, development, and distribution of the rodents as the most important group of vertebrates involved in rickettsial evolution.

REFERENCES

1. **Grandjean, F.**, Observations sur les acariens, *Bull. Mus. Hist. Nat. (l'ere Serie)*, 7, 119, 1935.
2. **Evans, G. O., Sheals, J. G., and MacFarlane, D.**, *The Terrestrial Acari of the British Isles*, Vol. 1, British Museum, London, 1961.
3. **Sharif, M.**, Historical review and relationships of the genera of the family Ixodidae, *Parasitology*, 26, 49, 1934.

4. **Clifford, C. M., Kohls, G. M., and Sonenshine, D. E.,** The systematics of the subfamily Ornithodorinae (Acarina: Argasidae). I. The genera and subgenera, *Ann. Entomol. Soc. Am.*, 57,429, 1964.
5. **Hoogstraal, H.,** personal communication.
6. **Ewing, H. E.,** The origin and classification of the trombiculid mites or Trombiculidae, *J. Wash. Acad. Sci.*, 39, 229, 1949.
7. **Hirst, S.,** On some arachnid remains from the Old Red Sandstone (Rhynie Chert Bed, Aberdeenshire), *Ann. Mag. Nat. Hist.*, 12, 455, 1923.
8. **Stromer, L.,** Oldest known terrestrial arachnids, *Science*, 164, 1276, 1969.
9. **Tiegs, O. W. and Manton, S. M.,** The evolution of the Arthropoda, *Biol. Rev.*, 33, 255, 1959.
10. **Colbert, E. H.,** *The Age of Reptiles*, W. W. Norton, New York, 1966, 1.
11. **Hoogstraal, H.,** Biology of ticks, in *Tick-borne Diseases and Their Vectors*, Wilde, J. K. H., Ed., Edinburgh University Press, Edinburgh, 1978, 3.
12. **Weidner, H.,** A tick, *Exodes succineus sp. n.*, found in baltic amber, *Veroeff. Uberseemus. Bremen, Reihe A*, 3, 143, 1964.
13. **Roberts, F. H. S.,** The Australian species of *Aponomma* and *Amblyomma* (Ixodoidea), *Aust. J. Zool.*, 1, 111, 1953.
14. **Gregory, W. K.,** *Evolution Emerging*, Vol. 1, Macmillan, New York, 1951, 1.
15. **Rehacek, J.,** Development of animal viruses and rickettsiae in ticks and mites, *Annu. Rev. Entomol.*, 10, 1, 1965.
16. **Roshidy, M. A.,** Observations by electron microscopy and other methods on the extra cellular rickettsia-like microorganisms of *Argas persicas* Oken (Ixodoidea, Argasidae), *J. Insect Pathol.*, 3, 148, 1961.
17. **Hoogstraal, H., Kaiser, M. N., and Kohls, G. M.,** The subgenus *Persicargas* (Ixodoidea, Argasidae, *Argas*). V. *Argas zumpti*, new species, a parasite of the South African cape vulture, *Ann. Entomol. Soc. Am.*, 61, 744, 1968.
18. **Steinhaus, E. A.,** *Principles of Insect Pathology*, McGraw-Hill, New York, 1949, 1.
19. **Buchner, P.,** Symbiose und Anpassung, *Nova Acta Leopold.*, 8, 257, 1940.
20. **Weiss, E.,** Family I. Rickettsiaceae, Tribe III. Wolbachieae Philip, 1955, in *Bergey's Manual of Determinative Bacteriology*, 8th ed., Buchanan, R. E. and Gibbons, N. E., Eds., Williams & Wilkins, Baltimore, 1974, 897.
21. **Hoogstraal, H.,** Ticks and spirochaetes, *Acta Trop.*, 36, 133, 1979.
22. **Davis, G. E.,** The spirochaetes, *Annu. Rev. Microbiol.*, 2, 305, 1948.
23. **Heisch, R. B.,** Studies in East African relapsing fever, *East Afr. Med. J.*, 27, 1, 1950.
24. **Varma, G. R.,** Transmission of relapsing fever spirochaetes by ticks. Aspects of disease transmission by ticks, *Symp. Zool. Soc. London*, 6, 61, 1962.
25. **Geigy, R.,** Relapsing fevers, in *Infectious Blood Diseases of Man and Animals*, Vol. 2, Weinman, D. and Ristic, M., Eds., Academic Press, New York, 1968, 175.
26. **Riek, R. F.,** Babesiosis, in *Infectious Blood Diseases of Man and Animals*, Vol. 2, Weinman, D. and Ristic, M., Eds., Academic Press, New York, 1968, 219.
27. **Shortt, H. E.,** Ticks and piroplasms. Aspects of disease transmission by ticks, *Symp. Zool. Soc. London*, 6, 157, 1962.
28. **Barnett, S. F.,** Theilerosis, in *Infectious Blood Diseases of Man and Animals*, Vol. 2, Weinman, D. and Ristic, M., Eds., Academic Press, New York, 1968, 269.
29. **Cheissin, E. M.,** The systematical position of piroplasmids and some peculiarities of their development in the vertebrate and invertebrate hosts, in *Proc. 1st Int. Congr. Parasitol., Rome*, Vol. 1, Pergamon Press, Oxford, 1964, 259.
30. **Gothe, R.,** Zur Entwicklung von *Aegyptianella pullorum* Carpano, 1928, in der Lederzecke *Argas (Persicargas) persicas* (Oken, 1818), *Z. Parasitenk.*, 29, 103, 1967.
31. **Bedford, H. A. H. and Coles, J. D. W. A.,** The transmission of *Aegyptianella pullorum* Carpano to fowls by means of ticks belonging to the genus *Argus*, *Onderstepoort J. Vet. Sci. Anim. Ind.*, 1, 15, 1933.
32. **Burgdorfer, W.,** *Ornithodoros moubata* als Testobjek bei Q-fieberfallen in der Schweg, *Acta Trop.*, 8, 41, 1951.
33. **Weyer, F.,** Die Beziehungen des Q-Fieber-Erregers (*Rickettsia burneti*) zu Arthropoden, *Z. Tropenmed. Parasitol.*, 4, 344, 1953.
34. **Weyer, F.,** Experimente mit einem Stamm von Zeckenbissfieber aus Nordqueensland, *Schweiz. Z. Allg. Pathol. Bakteriol.*, 22, 609, 1959.
35. **Kordova, N. and Rehacek, J.,** Microscopical investigations of organs from ticks infected with *Rickettsia prowazeki*, *Acta Virol Engl. Ed.*, 8, 465, 1964.

36. **Cowdry, E. V.,** Studies on the etiology of heartwater. II. *Rickettsia ruminantium* in the tissues of ticks transmitting the disease, *J. Exp. Med.*, 42, 253, 1925.
37. **Bertram, D. S.,** Rickettsial infections and ticks. Aspects of disease transmission by ticks, *Symp. Zool. Soc. London*, 6, 179, 1962.
38. **Hurlbut, H. S. and Thomas, J. I.,** The experimental host range of the arthropod-borne animal viruses in arthropods, *Virology*, 12, 391, 1960.
39. **Storey, H. H.,** The inheritance by an insect vector of the ability to transmit a plant virus, *Proc. R. Soc. London Ser. B*, 112, 46, 1932.
40. **Rochow, W. F.,** Barley yellow dwarf virus: phenotypic mixing and vector specificity, *Science*, 167, 875, 1970.
41. **Philip, C. B.,** Evidence of masking and latency in the fields of insect and helminth microbiology, in *Symposium on Latency and Masking in Viral and Rickettsial Infections*, Walker, D. L., Hanson, R. P., and Evans, A. S., Eds., Burgess Publishing, Minneapolis, 1958, 88.
42. **Smith D. J. W.,** Studies in the epidemiology of Q fever. III. The transmission of Q fever by the tick *Haemaphysalis humerosa, Aust. J. Exp. Biol. Med. Sci.*, 18, 103, 1940.
43. **Bozeman, E. M., Humphries J. W., and Campbell, J. M.,** A new group of rickettsia-like agents recovered from guinea pigs, *Acta Virol. Engl. Ed.*, 12, 87, 1968.
44. **Pope, J. H., Cook, I., and Campbell, R. W.,** A rickettsia-like organism from a water rat, *Hydromys chrysogaster*, from Brisbane, *Med. J. Aust.*, 2, 1080, 1967.
45. **Storz, J.,** Uber eine Naturliche Infektion eines Meerschweinchenbestandes mit einem Erreger aus der Psittakose-Lymphogranuloma Gruppe, *Zentralbl. Bakteriol. Parasitenkd. Infectionskr. Hyg. Abt. 1 Orig.*, 193, 432, 1964.
46. **Murray, E. S.,** Guinea Pig inclusion conjunctivitis virus. I. Isolation and identification as a member of the psittacosis-lymphogranuloma-trachoma group, *J. Infect. Dis.*, 114, 1, 1964.
47. **Robinson, G. W.,** A naturally occurring latent *Bedsonia* infection in guinea pigs, *Br. Vet. J.*, 125, 23, 1969.
48. **Gerloff, R. R. and Watson, R. O.,** A chlamydia from the peritoneal cavity of mice, *Infect. Immun.*, 1, 64, 1970.
49. **Pimental, D.,** Population regulation and genetic feedback, *Science*, 159, 1432, 1968.
50. **Burgdorfer, W., Anacker, R. L., Bird, R. G., and Bertram, D. S.,** Intranuclear growth of *Rickettsia rickettsii, J. Bacteriol.*, 96, 1415, 1968.
51. **Sclater, P. L.,** On the general geographic distribution of the members of the class Aves, *J. Linn. Soc. London Zool.*, 2, 130, 1858.
52. **Wallace, A. R.,** *The Geographical Distribution of Animals*, Vols. 1 and 2, Macmillan, London, 1876, 1.
53. **Schmidt, K. P.,** Faunal realms, regions, and provinces, *Q. Rev. Biol.*, 29, 322, 1954.
54. **Clifford, C. M., Kohls, G. C., and Sonenshine, D. E.,** The systematics of the subfamily Ornithodorinae (Acarina: Argasidae). I. The genera and subgenera, *Ann. Entomol. Soc. Am.*, 57, 429, 1964.
55. **Hoogstraal, H., Clifford, C. M., and Keirans, J. E.,** *Argas (Microargas) transversus* (Ixodoidea: Argasidae) of Galapagos giant tortoises: description of the female and nymph, *Ann. Entomol. Soc. Am.*, 66, 727, 1973.
56. **Schulze, P.,** Bestimmungstabelle fur das Zechengenus *Hyalomma* Koch, *Setzungeb. Ges. Naturf. Freunde, Berlin, Jahrg*, 189, 1919.
57. **Nuttall, G. H. F. and Warburton, G.,** Part II. Ixodidae, in *A Monograph of the Ixodoidea*, Nuttall, H. F., Warburton, G., Cooper, W. F., and Robinson, L. E., Eds., Cambridge University Press, Cambridge, 1911, 105.
58. **Blanc, G., Ascione, L., and Besiat, P.,** Rickettsemie experimentale de *Testudo mauritanica* avec *Rickettsia burneti* et infection de la tique *Hyalomma aegyptium, Bull. Soc. Pathol. Exot.*, 52, 564, 1959.
59. **Blanc, G., Ascione, L., and Besiat, P.,** Rickettsiemie experimentale de *Testudo graeca* avec *Rickettsia burneti*. Longue persistance de l'infection chez la tortue, passage et conservation du virus chez la tique *Hyalomma aegyptium, Arch. Inst. Pasteur Maroc.*, 6, 49, 1960.
60. **Kazar, J., Rehacek, J., and Brezina, R.,** Cultivation of rickettsiae in colubrid snake tissue cultures, *J. Hyg. Epidemiol. Microbiol. Immunol.*, 10, 240, 1966.
61. **Schindarow, L.,** Die Zuchtung von *Rickettsia rickettsii* (Art *Dermacentroxenus conori*) in der Gewebekultur der Niere eins Kaltbluters (*Testudo graeca*), *Zentralbl. Bakteriol. Parasitenkd. Infektionskr. Hyg. Abt. 1 Orig.*, 202, 46, 1967.
62. **Hoogstraal, H.,** African Ixodoidea. I. Ticks of the Sudan, Department of the Navy, Bureau of Medicine and Surgery, Washington, D.C., 1956, 1.

63. **Robinson, L. E.,** Part IV. The Genus Amblyomma, in *A Monograph of the Ixodoidea*, Nuttall, H. F., Warburton, G., Cooper, W. F., and Robinson, L. E., Eds., Cambridge University Press, Cambridge, 1926, 1.
64. **Roberts, F. H. S.,** Further observations on the Australian species of *Aponomma* and *Amblyomma* with descriptions of the nymphs of *Amblyomma moreliae* (L. Koch) and *Amblyomma loculosum* Neumann (Acarina: Ixodidae), *Aust. J. Zool.*, 12, 288, 1964.
65. **Roberts, F. H. S.,** The larvae of Australian Ixodidae (Acarina: Ixodoidea), *J. Aust. Entomol. Soc.*, 8, 37, 1969.
66. **Cooley, R. A.,** The genera *Boophilus, Rhipicephalus* and *Haemaphysalis* (Ixodidae) of the New World, *U.S. Natl. Inst. Health Bull.*, 187, 1, 1946.
67. **Neitz, W. O.,** Classification, transmission and biology of piroplasms of domestic animals, *Ann. N.Y. Acad. Sci.*, 64, 56, 1956.
68. **Berge, T. O.,** International Catalogue of Arboviruses Including Certain Other Viruses of Vertebrates, 2nd ed., Publ. No. 75-8301, Center for Disease Control, U.S. Department of Health, Education, and Welfare, Atlanta, 1975, 636.
69. **Gear, J.,** The rickettsial diseases of southern Africa. A review of recent studies, *S. Afr. J. Clin. Sci.*, 5, 158, 1954.
70. **Hoogstraal, H. and Kaiser, M. N.,** *Boophilus kohlsi* n. sp. (Acarina: Ixodidae) from sheep and goats in Jordan, *J. Parasitol.*, 46, 441, 1960.
71. **Kaiser, M. N. and Hoogstraal, H.,** The *Hyalomma* ticks (Ixodoidea, Ixodidae) of Afghanistan, *J. Parasitol.*, 49, 130, 1963.
72. **Dhanda, V. and Rao, T. R.,** The status of *Rhipicephalus sanguineaus* (Latreille, 1806) and *Rhipicephalus turanicus* Pomerantzev, 1940 (Acarina: Ixodidae) in India, *J. Bombay Nat. Hist. Soc.*, 66, 211, 1969.
73. **Kaiser, M. N. and Hoogstraal, H.,** Noteworthy recent tick records from Egypt. II. Hosts, distribution and ecology of *Rhipicephalus simus* Koch, *J. Egypt. Public Health Assoc.*, 42, 231, 1967.
74. **Morel, P. C. and Vassiliades, G.,** Les *Rhipicephalus* du groupe *sanguineus:* especes Africaines (Acariens: Ixodoidea), *Rev. Elev. Med. Vet. Pays Trop. N.S.*, 15, 343, 1963.
75. **Hoogstraal, H., Lim, B. L., Nadchatram, M., and Anastos, G.,** The Gunong Benom Expedition, 1967. VIII. Ticks (Ixodidae) of Gunong Benom and their altitudinal distribution, hosts and medical relationships, *Bull. Br. Mus. Nat. Hist. Zool.*, 23, 165, 1972.
76. **Kohls, G. M.,** Ticks (Ixodoidea) of the Philippines, National Institute of Health Publ. No. 192, Washington, D.C., 1950, 1.
77. **Hoogstraal, H., Kohls, G. M., and Parrish, D. W.,** Studies on Southeast Asian *Haemaphysalis* ticks (Ixodoidea, Ixodidae). *H. (Kaiseriana) psalistos sp. n.*, a parasite of deer and boars in Luzon, Philippines, *J. Parasitol.*, 53, 1096, 1967.
78. **Hoogstraal, H., Kohls, G. M., and Parrish, D. W.,** Studies on Southeast Asian *Haemaphysalis* ticks (Ixodoidea, Ixodidae). *Haemaphysalis (Kaiseriana) susphilippensis sp. n.*, a parasite of Luzon and Mindanao boars, *J. Parasitol.*, 54, 616, 1968.
79. **Trapido, H., Varma, M. G. R., Rajagopalan, P. K., Singh, K. R. P., and Rebello, M. J.,** A guide to the identification of all stages of the *Haemaphysalis* ticks of South India, *Bull. Entomol. Res.*, 55(2), 249, 1964.
80. **Kohls, G.,** Records and new synonymy of New World *Haemaphysalis* ticks with descriptions of the nymph and larva of *H. juxtakochi* Cooley, *J. Parasitol.*, 46, 355, 1960.
81. **Roberts, F. H. S.,** A systematic study of the Australian species of the genus *Haemaphysalis* Koch (Acarina: Ixodidae), *Aust. J. Zool.*, 11, 35, 1963.
82. **Hoogstraal, H.,** Ticks (Ixodoidea) of the Malagasy Faunal Region (excepting the Seychelles). Their origins and host-relationships; with descritpions of five new *Haemaphysalis* species, *Bull. Mus. Comp. Zool. Harv. Univ.*, 111, 37, 1953.
83. **Millot, J.,** La faune Malgache et le mythe Gondwanien, *Mem. Inst. Sci. Madagascar Ser. A*, 7, 1, 1952.
84. **Fooden, J.,** Breakup of Pangaea and isolation of relict mammals in Australia, South America and Madagascar, *Science*, 175, 894, 1972.
85. **Dietz, R. S. and Holden, J. C.,** Reconstruction of Pangaea: breakup and dispersion of continents, Permian to Present, *J. Geophys. Res.*, 75, 4939, 1970.
86. **Pomerantzev, B. I.,** Fauna of the U.S.S.R. Arachnida, *Acad. Sci. U.S.S.R.*, 4, 1, 1950.
87. **Teng, K. F.,** General description of geographical distribution of ticks in China, Abstract of a paper presented at the Seminar on Acarology Entomol. Soc. China, Peking, 1963 (in Chinese); as cited in **Li, C.C. and Chang, N. K.,** *Acta Zootax. Sin.*, 1, 352, 1964.
88. **Hoogstraal, H.,** Phylogeny of *Haemaphysalis* ticks, *Proc. XII Int. Congr. Entomol.*, Treherne, J. E. and Beamont, J. W. L., Eds., Academic Press, London, 1965, 760.

89. **Hoogstraal, H. and McCarthy, V. C.,** Hosts and distribution of *Haemaphysalis kashmirensis* with description of immature states and definition of the subgenus *Herpetobia* Canestrini (restricted), *J. Parasitol.*, 51, 674, 1965.
90. **Hoogstraal, H. and Wilson N.,** Studies on Southeast Asian *Haemaphysalis* ticks (Ixodoidea, Ixodidae). *Haemaphysalis (Alloceraea) vietnamensis sp. n.*, the first structurally primitive haemaphysalid recorded from southern Asia, *J. Parasitol.*, 52, 614, 1966.
91. **Hoogstraal, H.,** Notes on African *Haemaphysalis* ticks. IV. Description of Egyptian populations of the yellow dog ticks, *Haemaphysalis leachii leachii* (Audouin, 1827) (Ixodoidea, Ixodidae), *J. Parasitol.*, 44, 548, 1958.
92. **Luh, P. L. and Woo, W. C.,** A list of Chinese ticks, *Chin. J. Entomol.*, 1, 195, 1959.
93. **Sartbayev, S. K.,** Ticks of the genus *Haemaphysalis* in Kirgizia, *Prir. Ochag. Bolez. Vop. Parasit. Akad. Nauk Kazakh S.S.R.*, 4, 484, 1961.
94. **Li, C. C. and Chang, N. K.,** A new record of *Haemaphysalis* from China. *Haemaphysalis aponommoides, Acta Zootax. Sin.*, 1, 352, 1964.
95. **Hoogstraal, H.,** *Haemaphysalis tibetensis sp. n.* and its significance in elucidating phylogenetic patterns in the genus (Ixodoidea, Ixodidae), *J. Parasitol.*, 51, 452, 1965.
96. **Hoogstraal, H.,** Studies on Southeast Asian *Haemaphysalis* ticks (Ixodoidea, Ixodidae). *H. (H.) capricornis sp. n.*, the large Malayan serow haemaphysalid from southwestern Thailand, *J. Parasitol.*, 52, 783, 1966.
97. **Hoogstraal, H.,** *Haemaphysalis (Allophysalis) pospelov astromae sp. n.* from U.S.S.R. and redescription of the type material of *H. (A.) warburtoni* Nuttall from China (Ixodoidea, Ixodidae), *J. Parasitol.*, 52, 787, 1966.
98. **Hoogstraal, H.,** *Haemaphysalis (Herpetobia) himalaya sp. n.* (Ixodoidea, Ixodidae), a parasite of the Himalayan thar (Artiodactyla, Caprinae) in northwestern India, *J. Parasitol.*, 52, 805, 1966.
99. **Hoogstraal, H. and Trapido, H.,** Redescription of the type materials of *Haemaphysalis (Kaiseriana) bispinosa* Neumann (India), *H. (K.) neumanni* Donitz (Japan), *H. (K.) lagrangei* Larrouse (Vietnam) (Ixodoidea, Ixodidae), *J. Parasitol.*, 52, 1188, 1966.
100. **Hoogstraal, H. and Trapido, H.,** Studies on Southeast Asian *Haemaphysalis* ticks (Ixodoidea, Ixodidae). Species described by Supino in 1897 from Burma, with special references to *H. (Rhipistoma) asiaticus (=H. dentipalpis* Warburton and Nuttall), *J. Parasitol.*, 52, 1172, 1966.
101. **Mitchell, C. J., Hoogstraal, H., Schaller, G. B., and Spillett, J.,** Ectoparasites from mammals in Kanha National Park, Madhya Pradesh, India, and their potential disease relationships, *J. Med. Entomol.*, 3, 113, 1966.
102. **Reznik, P. A.,** On distribution area features of ticks belonging to the genus *Haemaphysalis* in the Soviet fauna, *Tezisy Dokl. 1. Akarol. Soveshch.* Bykhovsky, B. E., Ed., Akademiya Nauk SSSR, Moscow, 1966, 171.
103. **Hoogstraal, H., Kohls, G. M., and Trapido, H.,** Studies on Southeast Asian *Haemaphysalis* ticks (Ixodoidea, Ixodidae). *H. (Kaiseriana) anomala* Warburton: redescription, hosts and distribution, *J. Parasitol.*, 53, 196, 1967.
104. **Hoogstraal, H. and Rack, G.,** Ticks (Ixodidae) collected by Deutsche Indien-expedition, 1955-1958, *J. Med. Entomol.*, 4, 284, 1967.
105. **Hoogstraal, H., Uilenberg, G., and Klein, J. M.,** *Haemaphysalis (Rhipistoma) anoplos sp. n.*, a spurless tick of the *elongata* group (Ixodoidea, Ixodidae) parasitizing *Nesomys rufus* (Rodentia) in Madagascar, *J. Parasitol.*, 53, 1103, 1967.
106. **Hoogstraal, H., Santana, F. J., and Van Peenen, P. F. D.,** Ticks (Ixodoidea) of Mt. Sontra, Danang, Republic of Vietnam, *Ann. Entomol. Soc. Am.*, 61, 722, 1968.
107. **Hoogstraal, H., Roberts, F. H. S., Kohls, G. M., and Tipton, V. J.,** Review of *Haemaphysalis (Kaiseriana) longicornis* Neumann (resurrected) of Australia, New Zealand, New Caledonia, Fiji, Japan, Korea, and northeastern China and U.S.S.R., and its parthenogenetic and bisexual populations (Ixodoidea, Ixodidae), *J. Parasitol.*, 54, 1197, 1968.
108. **Hoogstraal, H. and Anastos, G.,** Studies on Southeast Asian *Haemaphysalis* ticks (Ixodoidea, Ixodidae). Redescription of *H. (Kaiseriana) renschi* Schulze (resurrected), and its hosts and distribution in Indonesia, *J. Parasitol.*, 54, 1214, 1968.
109. **Hoogstraal, H. and Kohls, G. M.,** Studies on Southeast Asian *Haemaphysalis* ticks (Ixodoidea, Ixodidae). Redescription of type material of *H. (Rhipistoma) heinrichi* Schulze, parasite of ferret-badger in Burma and Vietnam, *J. Parasitol.*, 54, 1057, 1968.
110. **Hoogstraal, H. and Parrish, D. W.,** Studies on Southeast Asian *Haemaphysalis* ticks (Ixodoidea: Ixodidae). *H. (Kaiseriana) luzonensis sp. n.* from Philippine deer and boars, *J. Parasitol.*, 54, 402, 1968.
111. **Wilson, N., Hoogstraal, H., and Kohls, G. M.,** Studies on Southeast Asian *Haemaphysalis* ticks (Ixodoidea, Ixodidae). Redescription of *H. (Rhipistoma) bartelsi* Schulze (resurrected), the Indonesian flying squirrel haemaphysalid, *J. Parasitol.*, 54, 1223, 1968.

112. **Hoogstral, H.,** *Haemaphysalis (Alloceraea) kitaokai sp. n.* of Japan, and keys to species in the structurally primitive subgenus *Alloceraea* Schulze of Eurasia (Ixodoidea, Ixodidae), *J. Parasitol.*, 55, 211, 1969.
113. **Hoogstraal, H., Lim, B. L., and Anastos, G.,** *Haemaphysalis (Kaiseriana) bispinosa* Neumann (Ixodoidea: Ixodidae): evidence for consideration as an introduced species in the Malay Peninsula and Borneo, *J. Parasitol.*, 55, 1075, 1969.
114. **Hoogstraal, H. and Dhanda, V.,** *Haemaphysalis (H.) darjeeling sp. n.* a member of the *H. (H.) birmaniae* group (Ixodoidea, Ixodidae), parasitizing artiodactyl mammals in Himalayan forests of India and Thailand, *J. Parasitol.*, 56, 169, 1970.
115. **Hoogstraal, H., Dhanda, V., and Bhat, H. R.,** *Haemaphysalis (Kaiseriana) davisi sp. n.* (Ixodoidea: Ixodidae), A parasite of domestic and wild mammals in northeastern India, Sikkim and Burma, *J. Parasitol.*, 56, 588, 1970.
116. **Hoogstraal, H. and Morel, P. C.,** *Haemaphysalis (Rhipistoma) hispanica* Gil Collado, a parasite of the European rabbit, redescription of adults and description of immature states (Ixodoidea: Ixodidae), *J. Parasitol.*, 56, 813, 1970.
117. **Hoogstraal, H. and Yamaguti, N.,** *Haemaphysalis (H.) pentalagi* Pospelova-Shtrom, a parasite of the Japanese black rabbit: redescription of the male and descriptions of the female, nymph, and larva (Ixodoidea: Ixodidae), *J. Parasitol.*, 56, 367, 1970.
118. **Hoogstraal, H., Kaiser, M. N., and Mitchell, R. M.,** *Anomalohimalaya lama*, new genus and new species (Ixodoidea: Ixodidae), a tick parasitizing rodents, shrews and hares in the Tibetan Highland of Nepal, *Ann. Entomol. Soc. Am.*, 63, 1576, 1970.
119. **Dhanda, V., Hoogstraal, H., and Bhat, H. R.,** *Haemaphysalis (Kaiseriana) ramachandrai sp. n.* (Ixodiodea, Ixodidae), a parasite of man and domestic and wild mammals in northern India and Nepal, *J. Parasitol.*, 56, 823, 1970.
120. **Hoogstraal, H.,** Identity, distribution and hosts of *Haemaphysalis (Rhipistoma) indica* Warburton (resurrected) (Ixodoidea: Ixodidae), a carnivore parasite of the Indian Subregion, *J. Parasitol.*, 56, 1013, 1970.
121. **Hoogstraal, H.,** *Haemaphysalis (H.) birmaniae* Supino and *H. (H.) goral sp. n.* (Ixodoidea: Ixodidae), Asian parasites of mammals, *J. Parasitol.*, 56, 1227, 1970.
122. **Hoogstraal, H. and El Kammah, K. M.,** *Haemaphysalis (Herpetobia) himalaya* Hoogstraal (Ixodoidea: Ixodidae), description of immature stages, hosts and distribution, *J. Parasitol.*, 56, 1023, 1970.
123. **Dhanda, V. and Bhat, H. R.,** *Haemaphysalis (Herpetobia) sundrai* Sharif (Ixodoidea: Ixodidae), a tick parasitizing sheep in the western Himalayas, redescription of female, description of male and exological observations, *J. Parasitol.*, 57, 646, 1971.
124. **Hoogstraal, H.,** Identity, hosts and distribution of *Haemaphysalis (Rhipistoma) canestrinii* (Supino) (resurrected), the postulated Asian progenitor of the African *leachi* complex (Ixodoidea: Ixodidae), *J. Parasitol.*, 57, 161, 1971.
125. **Hoogstraal, H.,** *Haemaphysalis (H.) sambar sp. n.* (Ixodoidea: Ixodidae), a parasite of the sambar deer in southern India, *J. Parasitol.*, 57, 173, 1971.
126. **Hoogstraal, H.,** *Haemaphysalis (Allophysalis) warburtoni* Nuttall: description of immature stages, adult structural variation, and hosts and ecology in Nepal, with redefinition of the subgenus *Allophysalis* Hoogstraal (Ixodoidea: Ixodidae), *J. Parasitol.*, 57, 1083, 1971.
127. **Hoogstraal, H.,** *Haemaphysalis (Kaiseriana) borneata sp. n.* (Ixodoidea: Ixodidae), a tick of the *H. (K.) aculeata* group parasitizing the sambar deer in Borneo, *J. Parasitol.*, 57, 1096, 1971.
128. **Hoogstraal, H., Saito, Y., Dhanda, V., and Bhat, H. R.,** *Haemaphysalis (H.) obesa* Larrousse (Ixodoidea: Ixodidae) from Northeast India and Southeast Asia: description of immature stages and biological observations, *J. Parasitol.*, 57, 177, 1971.
129. **Hoogstraal, H., Walker, J. B., and Neitz, W. O.,** Notes on African *Haemaphysalis* ticks. VII. *H. (Rhipistoma) hyracophila sp. n.* (Ixodoidea: Ixodidae), a parasite of hyraxes in South Africa, *J. Parasitol.*, 57, 417, 1971.
130. **Hoogstraal, H. and El Kammah, K. M.,** Studies on Southeast Asian *Haemaphysalis* ticks (Ixodoidea: Ixodidae). *H. (H.) traubi* kohls, redescription of male, description of female and new artiodactyl host and Malayan distribution records, *J. Parasitol.*, 57, 426, 1971.
131. **Hoogstraal, H. and Mitchell, R. M.,** *Haemaphysalis (Alloceraea) aponommoides* Warburton (Ixodoidea: Ixodidae), description of immature stages, hosts, distribution, and ecology in India, Nepal, Sikkim and China, *J. Parasitol.*, 57, 635, 1971.
132. **Hoogstraal, H., Dhanda, V., and El Kammah, K. M.,** *Aborphysalis*, a new subgenus of Asian *Haemaphysalis* ticks; and identity, distribution, and hosts of *H. aborensis* Warburton (resurrected) (Ixodoidea: Ixodidae), *J. Parasitol.*, 57, 748, 1971.

133. **Hoogstraal, H., El Kammah, K. M., Kadarsan, S., and Anastos, G.,** *Haemaphysalis (H.) sumatraensis sp. n.* (Ixodoidea: Ixodidae), a tick parasitizing the tiger, boar and sambar deer in Indonesia, *J. Parasitol.*, 57, 1104, 1971.
134. **Saito, Y., Hoogstraal, H., and Vasuvat, C.,** Studies on Southeast Asian *Haemaphysalis* ticks (Ixodoidea: Ixodidae). Identity and hosts of immature states of *H. (Rhipistoma) asiatica* (Supino) from Thailand, *J. Parasitol.*, 57, 1099, 1971.
135. **Hoogstraal, H., Dhanda, V., and Bhat, H. R.,** *Haemaphysalis (Kaiseriana) anomala* Warburton (Ixodoidea: Ixodidae) from India: description of immature stages and biological observations, *J. Parasitol.*, 58, 605, 1972.
136. **Hoogstraal, H.,** Redescription of the type material of *Haemaphysalis (Aboimisalis) cinnabarina* (revalidated) and its junior synonym *H. (A.) sanguinolenta* described by Koch in 1844 from Brazil (Ixodoidea: Ixodidae), *J. Parasitol.*, 59, 379, 1973.
137. **Hoogstraal, H., Garney, W. P., Kadarsan, S., and Van Peenen, P. F.,** *Haemaphysalis (Kaiseriana) celebensis* Hoogstraal, Trapido and Kohls (Ixodoidea: Ixodidae), a Walacean member of the *hystricis* group; identity, distribution, hosts and ecology, *J. Parasitol.*, 59, 556, 1973.
138. **Hoogstraal, H. and Wassef, H. Y.,** The *Haemaphysalis* ticks (Ixodoidea: Ixodidae) of birds. III. *H. (Ornithophysalis) subgen. n.*: definition of species, hosts, and distribution in the Oriental, Palearctic, Malagasy and Ethiopian Faunal Regions, *J. Parasitol.*, 59, 1099, 1973.
139. **Hoogstraal, H., El Kammah, K. M., Santana, F. J., and Van Peenen, P. F. D.,** Studies of Southeast Asian *Haemaphysalis* ticks (Ixodoidea: Ixodidae). *H. (Kaiseriana) lagrangei* Larrousse: identity, distribution and hosts, *J. Parasitol.*, 59, 1118, 1973.
140. **Emel'yanova, N. D. and Hoogstraal, H.,** *Haemaphysalis verticalis* Itagaki, Noda and Yamaguchi: rediscovery in China, adult and immature identity, rodent hosts, distribution, and medical relationships (Ixodoidea: Ixodidae), *J. Parasitol.*, 59, 724, 1973.
141. **Saito, Y. and Hoogstraal, H.,** *Haemaphysalis (Kaiseriana) megeshimaensis* sp. n. (Ixodoidea: Ixodidae), a Japanese deer parasite with bisexual and parthenogenetic reproduction, *J. Parasitol.*, 59, 569, 1973.
142. **Cerny, V. and Hoogstraal, H.,** *Haemaphysalis (Allophysalis) danieli* sp. n. (Ixodoidea: Ixodidae), female and tentatively associated immature stages from high mountains of northern Pakistan and Afghanistan, *J. Parasitol.*, 63, 567, 1977.
143. **Hoogstraal, H. and Wassef, H. Y.,** *Haemaphysalis (Allophysalis) kopetdaghica:* identity and discovery of each feeding stage on the wild goat in northern Iran (Ixodoidea: Ixodidae), *J. Parasitol.*, 65, 783, 1979.
144. **Marchette, N. J.,** Rickettsioses (tick typhus, Q-fever, urban typhus) in Malaya, *J. Med. Entomol.*, 2, 339, 1965.
145. **Roshidy, M. A.,** Structure of the female reproductive system of *Ixodes ricinus* (L.), and its bearing on the affinity of *Ixodes* to other ixodid genera, *J. Parasitol.*, 55, 1078, 1969.
146. **Kahn, J.,** Cytotaxonomy of ticks, *Q. J. Microsc. Sci.*, 105, 123, 1964.
147. **Arthur, D. R.,** *Ticks of the Genus Ixodes in Africa*, Athlone Press, London, 1965, 1.
148. **Cooley, R. A. and Kohls, G. M.,** The genus *Ixodes* in North America, Bull. No. 184, U.S. National Institute of Health, Washington, D.C., 1945, 1.
149. **Keirans, J. E. and Clifford, C. M.,** The genus *Ixodes* in the United States: a scanning electron microscope study and key to the adults, *J. Med. Entomol.*, Suppl. 2, 1, 1978.
150. **Anastos, G.,** The Ticks or Ixodides of the U.S.S.R., Publ. No. 548, U.S. Public Health Service, Washington, D.C., 1957, 1.
151. **Roberts, F.H.S.,** A systematic study of the Australian species of the genus *Ixodes, Aust. J. Zool.*, 8, 392, 1960.
152. **Theiler, G.,** Ticks in the South African Zoological Survey Collection. I. *Ixodes alloudi* Neumann 1913, a primitive tick parasite on shrews, *Onderstepoort J. Vet. Sci. Anim. Ind.*, 17, 51, 1941.
153. **MacCleod, J. and Gordon, W. S.,** Studies in louping-ill (an encephalomyelitis of sheep). II. Transmission by the sheep tick, *Ixodes ricinus* L., *J. Comp. Pathol.*, 45, 240, 1932.
154. **Smith, C. E. G.,** A virus resembling Russian spring summer encephalitis virus from an ixodid tick in Malaya, *Nature (London)*, 178, 581, 1956.
155. **Bancroft, W. H., Scott, R. M., Snitbalm, C., Weaver, R. E., Jr., and Gould, D. S.,** Isolation of Langat virus from *Haemaphysalis papuana* Thorell in Thailand, *Am. J. Trop. Med. Hyg.*, 25, 500, 1976.
156. **Giroud, P., Capponi, M., Dumas, N., and Rageau, J.,** Les *Ixodes ricinus* et *hexagonus* de France contienent des agents rickettsiens au proches, *C. R. Acad. Sci.*, 260, 4874, 1965.
157. **Domrow, R. and Derrick, E. H.,** *Ixodes holocyclus*, the man-biting tick in southeast Queensland, *Aust. J. Sci.*, 27, 234, 1965.

158. **Campbell, R. W. and Domrow, R.,** Rickettsiosis in Australia: isolation of *Rickettsia tsutsugamushi* and *R. australis* from naturally infected arthropods, *Trans. R. Soc. Trop. Med. Hyg.*, 68, 397, 1974.
159. **Zdrodovskii, P. F. and Golinevich, H. M.,** *The Rickettsial Diseases*, Pergamon Press, New York, 1960, 1.
160. **Hoogstraal, H., Kaiser, M. N., Traylor, M. A., Gaber, S., and Guindy, E.,** Ticks (Ixodoidea) on birds migrating from Africa to Europe and Asia, *Bull. W.H.O.*, 24, 197, 1961.
161. **Hoogstraal, H., Kaiser, M. N., Traylor, M. A., Guindy, E., and Gaber, S.,** Ticks ((Ixodidae) on birds migrating from Europe and Asia to Africa, 1959–61, *Bull. W.H.O.*, 28, 235, 1963.
162. **Nuorteva, P. and Hoogstraal, H.,** The incidence of ticks (Ixodoidea, Ixodidae) on migrating birds arriving in Finland during the Spring of 1962, *Ann. Med. Exp. Biol. Fenn.*, 41, 457, 1963.

Chapter 4

EVOLUTION AND DISTRIBUTION OF RODENTS

N. Marchette

TABLE OF CONTENTS

I. Introduction 54

II. Geological Evolution of the Earth 54
A. Plate Tectonics and Seafloor Spreading 54
B. Breakup of Pangaea 55

III. Evolution of the Rodentia 56
A. Center of Origin 56
B. Time of Origin 57

IV. Classification and Distribution of the Rodentia 58
A. Hystricomorpha 58
B. Sciuromorpha 61
C. Myomorpha 63
1. Australian Rodent Fauna 63
2. Rodent Fauna of the Oriental Region 65
a. India 65
b. Philippines 67
c. Southeast Asia 67

V. Summary 68

References 70

I. INTRODUCTION

Rodents are hosts of many ixodid ticks, which are themselves the natural hosts of rickettsiae. The aim of this chapter is to review briefly the evolution of the major rodent taxa and their present distribution as important factors in rickettsial evolution and distribution. Other vertebrates, especially large herbivores, are also hosts of many ticks and undoubtedly have played a prominent role in the evolution of some rickettsial genera, but rodents are the primary vertebrate hosts of present members of the genus *Rickettsia* or their immediate ancestors. Since the evolution of *Rickettsia* is our primary concern, a brief discussion of rodent evolution and distribution should be useful. No attempt will be made to critically review the extensive literature on this subject, but the references cited should provide the interested reader with the sources for a comprehensive study.

II. GEOLOGICAL EVOLUTION OF THE EARTH

The basic pattern of vertebrate evolution and distribution and that of their parasites was greatly influenced, if not wholly determined, by events of incalculable magnitude that occurred near the beginning of the Mesozoic, some 225 million years ago. Contrary to earlier opinions, the earth is not a static sphere of continents and oceans permanently fixed in place since the beginning. It is now accepted dogma that the continents are, and always have been, in continuous movement, drifting, as it were, like enormous icebergs floating on a sea of dense magma beneath. Similar to icebergs, the continents in the distant past have drifted together, colliding to form new supercontinents and then, breaking up in new configurations, drifted apart again.[1]

A. Plate Tectonics and Seafloor Spreading

During the Paleozoic, an era that lasted nearly 350 million years, six major continents (Gondwana, Laurentia, Baltica, Siberia, Kazakhstania, and China) drifted together on a collision course eventually to form the super supercontinent of Pangaea in the late Permian.[1-3] At the beginning of the Mesozoic, near the end of the Triassic Period, the process of continental fusion was reversed and rifting began. The first suggestion, based on geological evidence that land masses must have moved bodily over great distances, was presented by Taylor,[4] but Wegener[5] seems to have been the first to point out that if the South American and African continents were "pushed" together, the eastern edge of the former would fit neatly into the western portion of the latter. However, continents cannot be so easily "pushed" around and the idea fell into disrepute for lack of a plausible mechanism that might generate the enormous forces necessary to accomplish it. It wasn't until 40 years later that Hess[6] suggested the mechanism by which continental drift operates—a process named seafloor spreading[7] and now called plate tectonics.[8] A satisfactory treatment of plate tectonics is beyond the scope of this work, but technical and semitechnical accounts are readily available for those interested.[2,3,9-16] Perhaps the most conclusive evidence for seafloor spreading and continental drift was the discovery of the regular pattern of magnetic anomalies in the deep ocean rocks.[17,18] This led to the development of the theory of plate tectonics.[19-21] Supporting evidence for the theory was provided by the distribution of seismically active regions of the earth. Most earthquakes are confined to narrow belts which outline the plate boundaries.[8]

Earlier, the disjunct distribution of a wide variety of Southern Hemisphere elements also suggested that the southern continents were once joined together.[22] Keast[23,24] critically reviewed this evidence and found it somewhat less than convincing, but later changed his mind.[25] The present distribution of characoid fishes, which occur only in

South America and Africa,[26] provided the best faunal evidence that the continents were once joined. These fish are strictly freshwater forms and could not have crossed a salt-water barrier. Darlington[27] noted and documented the peculiar distribution of carabid beetles, peloridiid bugs, *Nothofagus* trees, mammals, and fishes in the Southern Hemisphere, suggesting origin on a supercontinent with subsequent rifting and drifting apart. However, he was skeptical and concluded that the present distribution of plants and animals did not establish beyond a reasonable doubt the validity of the hypothesis of continental drift. The accumulated work of other investigators has shown his doubts to be unfounded.[25,28–43]

The relative positions occupied by our modern continents in Pangaea have been determined with considerable accuracy on the basis of contour fitting of the coasts and continental shelves,[2,44–51] stratigraphic geology,[50,52–58] and by paleomagnetism.[21,59–63] This is not by any means an exhaustive list of important and pertinent references to the large literature on continental drift, sea floor spreading, and plate tectonics. It is merely intended to provide the documentation for these concepts by modern pioneers in the field. Much has been written since by these and other investigators. Excellent general and popularized discussions of these phenomena are available.[15,64–69]

B. Breakup of Pangaea

In the Cambrian, the continents were distributed along the equator, but as they came together and fused in the Permian, they formed Pangaea stretching across the equator from pole to pole.[1] In the middle of the Triassic, about 200 million years ago, Pangaea began to break up with two rifts that resulted in the opening of the Atlantic and Indian Oceans by the end of the period.[70] The northern rift split Pangaea along an East-West line north of the Equator, creating Laurasia composed of North America and Eurasia in the North and Gondwanaland in the South. At about the same time, the southern rift split India, Australia, and Antarctica from the African-South American land mass. Almost immediately India was split from Antarctica or Australia[71,72] and began its long drift northward.

During the Jurassic, the major north and south blocks drifted apart and the North American portion rotated opening the Tethys Sea (named for the wife of Oceanus, who was the mother of the seas in Greek mythology), forerunner of the Mediterranean. By the end of the period, South America began splitting away from Africa,[73–75] and over the next 70 million years the continents slowly inched apart until the separation was complete at the end of the Cretaceous some 65 million years ago.

The Australia-Antarctica mass slowly rotated westward. Although rifting between these two continents probably began in the Cretaceous,[48] they did not drift apart until the Eocene with Australia moving north and Antarctica south.[21,39,63,76–79] During the Cretaceous, the two Americas were joined at the Isthmus of Panama, but a continuous land bridge probably did not form before the Miocene or Pliocene.[80] India completed its northward migration, colliding with Asia during the late Paleocene and early Eocene.[81] The enormous forces generated by collision of these huge land masses eventually uplifted the Himalayan Mountain chain. Madagascar is thought to have remained attached to Africa until the end of the Cretaceous,[39] but its unique, mostly Asian, fauna suggests an earlier split, perhaps at the end of the Permian or beginning of the Triassic.[82] The continents continued drifting throughout the Cenozoic (65 million years ago to the present) until they reached their present positions. They are still drifting, but at an infinitesimally slow rate.

Evolution of vertebrates—The breakup of Pangaea set the stage for the evolution of mammals and their parasites. By the time the continents had more or less assumed their present configurations and were drifting apart, the reptiles were no longer the dominant vertebrates. At the beginning of the Cenozoic, mammals and birds became, as they are today, the dominant terrestrial vertebrates. Fishes continue to dominate the seas, but, with one or two possible exceptions, have played no role in rickettsial evolution.

III. EVOLUTION OF THE RODENTIA

A. Center of Origin

Rodents are generally thought to have originated in the Palearctic Region, somewhere in what is now North or Central Asia. However, Darlington[83] presents convincing arguments (primarily on the basis of area and climate) for the Old World Tropics, rather than the North Temperate Zone, as the center of evolution of dominant vertebrates. The differences of opinion about the geographic origin of the Rodentia may be more apparent than real, since the present tropical belt is unusually narrow.[84] During the Eocene it extended at least 50° north and south of the equator.[85,86] The region between the 50th parallels of latitude includes southern Europe, all of China and the whole of Africa, much of which are now in the temperate zone. This vast area of warm, humid climate, coinciding in time with the tremendous development, divergence, and distribution of the rodents, lends credence to Darlington's hypothesis.

In Dobzhansky's opinion,[87] the processes of adaptation for life in the temperate and cold zones are primarily devoted to coping with the physical environment and obtaining food. In the tropics, although the physical environment is agreeable and food easily obtained, the biological environment is harsh and exacting. To quote Dobzhansky,

> The processes of natural selection which arise from encounters between living things and physical forces in their environment are different from those which stem from competition within a complex community of organisms. The struggle for existence in habitats in which harsh physical conditions are the limiting factors is likely to be rather passive as far as the organism is concerned.*

Abrupt change in physical conditions (such as cold) is likely to wipe out large numbers of organisms without respect for their individual characteristics except for those directly involved in resistance to the particular physical factors. Natural selection based on such limited traits often leads into evolutionary blind alleys.

In the tropics, the paramount adaptive problems become the interrelationships between competing and symbiotic species. To quote Dobzhansky again,

> The effectiveness of natural selection is by no means proportional to the severity of the struggle for existence. On the contrary, selection is most effective when, instead of more or less random destruction of masses of organisms, the survival and elimination acquire a different character. Individuals that survive and reproduce are mostly those that possess combinations of traits which make them attuned to the manifold reciprocal dependencies in the organic community . . . Tropical environments provide more evolutionary challenges than do environments of temperate or cold lands.*

The present tropical areas of the globe have probably undergone much less drastic climatic and biotic changes, even during the Pleistocene Ice Ages, than have temperate regions.

Thus one is tempted to place the actual origin of the Rodentia geographically in what are now, and have been for eons, the tropics of Asia and Africa. The world-wide climatic changes occurring at the end of the Mesozoic and beginning of the Cenozoic and extending the tropics far into what are now temperate zones would have provided the

*Dobzhansky, T., Evolution in the Tropics, *Am. Sci.*, 38, 209, 1950. Reprinted by permission, *American Scientist*, journal of Sigma Xi, The Scientific Research Society of North America.

necessary large area for the great dispersion and divergence of mammals that took place at that time.

The earliest known rodents, all belonging to the extinct family Paramyidae, appear only in late Paleocene deposits of North America and early Eocene deposits in Europe,[88] but this in no way excludes the vast tropical areas of Asia and Africa as the probable birthplace of the Rodentia. The fossil record of the Old World Tropics (those of Asia in particular) is poorly known in comparison to the great amount of paleontologic data available for the North Temperate Zone. The major mammalian fossil record extends back to the beginning of the Tertiary Period about 60 million years ago, but the record is temporally and geographically incomplete. Some periods and some areas have been more intensively studied than others—facts which may not be entirely unrelated.

The Neotropical area (including Mexico, the West Indies, Central and South America) has been called the headquarters of the Rodents based on the number of named forms found there.[89] All 3 suborders and 7 of Simpson's 15 superfamilies are represented in the region. However, to consider the Neotropics as the "headquarters" of the rodents is misleading if this is interpreted as denoting broad evolutionary significance. Almost half the 1500 named forms belong to the family Muridae and many others to Sciuridae. These two families do not appear in the fossil record of South America until the late Miocene or early Pliocene when the southern continent was connected to North America, from which it had been isolated by sea barriers since the beginning of the Tertiary.[90] Thus, geologically speaking they are recent immigrants to South America and certainly did not evolve there. Until well into the Pliocene, the Central American fauna was almost exclusively derived from that of North America. The geographical boundaries of the Neotropical Region are thus defensible only on the basis of recent faunal distribution.[91,92] This, of course, is the criterion on which the 19th century naturalists based their delineations of major faunal regions.

B. Time of Origin

Most mammalian orders already existed by the end of the Paleocene, indicating a somewhat earlier origin for many, if not all, of them. The stratigraphic record, such as it is, suggests a post-Cretaceous origin for most placental orders except insectivores, which arose earlier, but many must have had their origin in the late Cretaceous.[93] Despite the scanty fossil record, Wilson[93] concludes that the Rodentia differentiated from generalized insectivore stock present in the early to middle Paleocene, and by the late Paleocene they had acquired the fundamental structures of the order.

The development, divergence, and dispersion of Rodentia in the Eocene was so great that this epoch might be called the "Age of Rodents." Whether they originated in the tropics of Asia or Africa or both, the rodents soon spread over the Asian, African, and European continents. During the early and again in the late Eocene, when land bridges existed between northwestern North America and northeastern Asia, there was intense migration of animals between the Old and the New Worlds.[94] Species migrations in both directions via Bering Sea land bridges also occurred in the Oligocene, Miocene, Pliocene, and Pleistocene.[95] The earliest migrations brought the first rodents into North America; later migrations added new species to the developing New World fauna and brought New World species back into Asia. Late Mesozoic and early Cenozoic land connections between North America and Europe[96] also may have allowed the immigration of European and African faunal elements. Similar South America–Antarctica–Australia land bridges[97,98] permitted the migration of marsupials from South America into Australia. Placental mammals apparently did not reach South America by migration from North America until long after the continents had severed connections and drifted far apart. The first placental mammals (rodents of the family Muridae) did not reach Australia from Southeast Asia until much later and marsupials still are the dominant native animals.

The rodents now are well represented on all the great land masses of the world and have been major components of the mammalian fauna of all the continents (except Australia) for a very long time. The coincidental distribution of ixodid ticks and rickettsiae with that of certain rodent groups should come as no surprise. If, as suggested, ixodid ticks and rickettsiae (Genus *Rickettsia*) developed mainly in conjunction with rodents, their past distribution should have coincided with that of the rodents. This will be discussed later in greater detail. For the time being, it is sufficient to summarize what is known of the past and present distribution of rodents, doing this only in the most general terms by comparing the present distribution of living members of major rodent taxa with the distribution of fossil rodents.

IV. CLASSIFICATION AND DISTRIBUTION OF THE RODENTIA

In terms of number, distribution, and diversity of species, the rodents are dominant vertebrates. They now comprise a huge order containing about two thirds of all living mammal species, and evidence from the fossil record suggests that, since the mid-Eocene or earlier, they have always formed a major portion of the total mammalian fauna.[99] Several classifications of the Order have been proposed,[89,100–102] but that of Simpson[102] will be followed here (Table 1). He divided the Order Rodentia into three suborders: (1) the Sciuromorpha, most of which probably arose in the Nearctic Region, (2) the Myomorpha, predominantly of Palearctic origin, and (3) the Hystricomorpha, clearly Neotropical in origin and development. These major groups are comprised of 15 superfamilies and 32 families, only a few of which are known to have played any role in rickettsial evolution. The least in importance appear to have been the South American rodents.

A. Hystricomorpha

South America is currently the center of development of the Suborder Hystricomorpha with 11 of the 16 living families represented in the fauna compared to 5 in the rest of the world (Table 2). It has a unique fauna and is almost certainly the center of origin of the Cavioidea and Chinchilloidea and probably the evolutionary center of the Suborder Hystricomorpha. Outside South America and Central America, the hystricomorphs are represented currently by only the Hystricoidea distributed throughout the Old World Tropics including most of Africa and southern Palearctic Asia, the Bathyergoidea in sub–Saharan Africa, the Ctenodactyloidea in northern Africa, and one family of Octodontoidea south of the Sahara in Central and East Africa. One genus of the Erethizontoidea is found in North America, but its migration from South America is recent, probably no earlier than the upper Pliocene.

The distribution of fossil hystricomorphs is even more impressive. Of the 17 families recognized in the fossil record, 12 are found in South America; 3 are known only from Eurasia, 2 from Africa (Table 3). The two families in the North American fossil record, Erethizontidae and Hydrochoeridae, are found only in relatively recent strata and represent migrants from South America.[103,104]

Thus, the present distribution of living and fossil hystricomorphs suggests that their ancestors evolved on a land mass composed of the present continents of Africa and South America joined together. The similarities of New World and Old World monkeys further support this hypothesis, although a direct relationship between them is not universally accepted.[105] Wood,[88] Simpson,[90] Gidley,[106] and Cook,[107] among others, suggest that the evidence, scant though it may be, indicates that the South American hystri-

Table 1
THE LIVING FAMILIES OF RODENTS (ORDER RODENTIA) AND THEIR PROBABLE REGIONS OF ORIGIN

Name	Present distribution	Probable region of origin
	Suborder Sciuromorpha	
Superfamily Aplodontoidea		
Family Aplodontidae	North America	North America
Superfamily Sciuroidea		
Family Sciuridae	Worldwide (except Australia)	North America
Superfamily Geomyoidea		
Geomyidae	North America	North America
Family Heteromyidae	North America, Northern South America	North America
Superfamily Castoroidea		
Family Castoridae	Asia, North America, Europe	North America Eurasia
Superfamily Anomaluroidea		
Family Anomaluridae	Africa	Africa/Europe
Family Pedetidae	Africa	Africa
	Suborder Myomorpha	
Superfamily Muroidea		
Family Cricetidae	Europe, Asia, North and South America, Africa	North America/ Eurasia
Family Spalacidae	Europe, Africa, West Asia	Eurasia
Family Rhizomyidae	Asia, Africa	Eurasia
Family Muridae	Worldwide	Eurasia
Superfamily Gliroidea		
Family Gliridae	Europe, Asia, Africa	Eurasia
Family Platacanthomyidae	Asia	South China/India
Family Seleviniidae	Asia	Asia
Superfamily Dipodoidea		
Family Zapodidae	Europe, Asia, North America	Eurasia/North America
Family Dipodidae	Europe, Asia, Africa	Eurasia
	Suborder Hystricomorpha	
Superfamily Histricoidea		
Family Hystricidae	Europe, Asia, Africa	Indo-Malayan
Superfamily Erethizontoidea		
Family Erethizontidae	North and South America	South America
Superfamily Cavioidea		
Family Caviidae	South America	South America
Family Hydrochoeridae	South America	South America
Family Dinomyidae	South America	South America
Family Dasyproctidae	South America	South America
Superfamily Chinchilloidea		
Family Chinchillidae	South America	South America
Family Capromyidae	South America	South America

Table 1 (continued)
THE LIVING FAMILIES OF RODENTS (ORDER RODENTIA) AND THEIR PROBABLE REGIONS OF ORIGIN

Name	Present distribution	Probable region of origin
Superfamily Octodontoidea		
Family Octodontidae	South America	South America
Family Ctenomyidae	South America	South America
Family Abrocomidae	South America	South America
Family Echimyidae	South America	South America
Family Thryonomyidae	Africa	Africa
Family Petromyidae	Africa	Africa
Superfamily Bathyergoidea		
Family Bathyergidae	Africa	Africa
Superfamily Ctenodactyloidea		
Family Ctenodactylidae	Africa	Asia

Table 2
DISTRIBUTION OF LIVING FAMILIES OF RODENT SUPERFAMILIES

Order and superfamily	Palearctic	Ethiopian	Oriental	Nearctic	Neotropic	Australia	Total no. of families
Sciuromorpha							
Sciuroidea	1	1	1	1	1		1
Aplodontoidea				1			1
Geomyoidea				2	1		2
Castoroidea	1			1			1
Anomaluroidea		2					2
Myomorpha							
Muroidea	4	4	2	2	1	1	4
Glirioidea	3	1	1				3
Dipodoidea	1	1		1			
Hystricomorpha							
Hystricoidea	1	1	1				1
Erethyzontoidea				1	1		1
Octodontoidea		2			5		7
Cavioidea					4		4
Chinchilloidea					1		1
Bathyergoidea		1					1
Ctenodactyloidea		1					1
Total number of families	11	14	5	9	14	1	30

comorphs (and primates) evolved in North America and later migrated by island hopping into the southern continent. The northern forms then became extinct, possibly due to pressure from immigrating Asian species. The African hystricomorphs and primates arose independently, and their present similarity with South American forms is a result of parallel evolution. Our present knowledge of continental drift and plate tectonics, however, suggests that the South American hystricomorph fauna arose in Africa, or more accurately in Gondwanaland, rather than in North America.

The break-up of this giant land mass and subsequent drifting apart of Africa and South America could have begun as early as the end of the Cambrian[108] or the end of the Permian,[109] but the continents probably had not started drifting apart until 150 to 200 million years ago in the late Triassic to mid-Jurassic.[110] Hurley[111] suggested that the rift between West Africa and North America began in mid-Triassic, but the sepa-

Table 3
DISTRIBUTION OF FOSSIL FAMILIES OF RODENT SUPERFAMILIES

Order and superfamily	Eurasia	Ethiopian	Oriental	Nearctic	Neotropic	Australia	Total number
Sciuromorpha							
Aplodontoidea	3			5			5
Sciuroidea	1			1			1
Geomyoidea				2			2
Castoroidea	1			1			2
Anamaluroidea	2	2					4
Myomorpha							
Muroidea	4	1		1	1	1	4
Glirioidea	1						
Dipodoidea	2			2			4
Hystricomorpha							
Hystricoidea	1		1[a]				1
Erethizontoidea				1	1		1
Cavioidea				1	5		5
Chinchilloidea					1		1
Octodontoidea	1	1			5		7
Bathyergoidea	1					1	
Ctenodactyloidea	1		1[a]				1
Total number of families	17	5	2	14	13	1	39

[a]India.

ration of southern Africa from South America may not have occurred until the Cretaceous. Whatever date is accepted, it is clear the process began long before the differentiation of rodents from their generalized mesethmoid ancestors. The earliest known fossil hystricomorphs in either Africa or South America date back only to the Oligocene. The cataclysmic processes of fracturing and drifting apart of enormous land masses, however, occurred over many tens of millions of years. During this long period of geologic activity, many islands may have existed between the two continents allowing for the interchange of faunal elements. The present mid-Atlantic Ridge and Rift Valley may be the remnants of such a system of islands which could have served as stepping stones for the migration of ancestral rodent and simian elements between South America and Africa. The time factor is such, however, that it is unlikely that either hystricomorph rodents or primates had differentiated until the two great land masses were separated by a nearly uncrossable sea barrier. If this is true, then one must concede that the hystricomorph rodents and simians of Africa and South America actually developed along parallel lines but originated from common generalized ancestral stocks originating in the Southern Hemisphere. An alternative hypothesis suggests an African origin for caviamorph rodents, which reached South America by rafting[112] or possibly by migration to Europe and across the early Cenozoic land bridge to North America,[113,114] then south to the southern continent. The virtual isolation of South America from other major land masses for much of the Cenozoic provided the early hystricomorphs an unparalleled opportunity for development and divergence.

B. Sciuromorpha

The Sciuromorpha is a large and diverse order of seven distinct families (Table 1) whose geographical center of origin can be established only in very general terms.

Based on the number of superfamilies and families represented in the fauna, North America is now the home of the suborder. Only the Sciuridae is widespread outside the Nearctic Region, but it is a large group with well-developed components in the Palearctic, Oriental, Ethiopian, and Neotropical Regions. Fossil sciurids are known only from the Palearctic and Nearctic Regions, indicating that they have undergone major divergence and geographic spread in relatively recent geologic epochs. The fossil record also shows that the Sciuromorpha were formerly more widespread throughout Eurasia. The distribution of living and fossil forms (Tables 2 and 3) suggests that they are derived from early Palearctic and Nearctic ancestors. From what is known of the geology and paleoclimatology of Eurasia and North America, it seems likely that Laurasia was their original home with early extensions and isolation of one group. The Anomaluroidea is found only in Africa, but it may have arisen in Europe during the mid-Eocene or earlier and migrated South much later, subsequently dying out in Europe and now persisting only in Africa. At least one superfamily, the Geomyoidea, differentiated in the Nearctic during the Oligocene or earlier. They are now a major component of the North American fauna and are found nowhere else except in South America to which they migrated in recent times.

The Sciuridae is the only family in the order that is distributed throughout the world (except in Australia). It is especially well developed in the Nearctic, Palearctic, and Indo-Malayan regions. The earliest fossil sciurids are from the early Oligocene of North America, and Black[115] proposes a Nearctic origin with migration into Eurasia and spread to Western Europe beginning in the Middle Oligocene. There were other exchanges between Nearctic and Palearctic sciurids in later ages. Interestingly, arboreal forms predominate in the tropics, whereas most Palearctic and Nearctic genera are ground squirrels. However, flying squirrels are the predominant fossil forms in Europe,[116] and evidence from their ectoparasitic fleas[117] suggests that the tropical forms originated in the Palearctic. There are virtually no fossil flying squirrels in North America before the Pleistocene, suggesting that the present representatives in the Nearctic are derived from recent immigrants from Asia. This has an important bearing on the origin of *Rickettsia prowazekii* in *Glaucomys* in the eastern U.S. and will be discussed further in Volume I, Chapter 6.

The Castoridae appear in Oligocene deposits in Europe and North America. They are presently distributed over the Palearctic and Nearctic Regions, although only a single living genus is recognized.[89] The Castoroidea are usually considered to have evolved in the North Temperate Zone since no tropical forms are known.

The other two families in the Order are small and highly restricted geographically. Living members of the Aplodontidae are found only in the northwest corner of the U.S. and southwestern extreme of British Columbia. Their fossil record extends all the way back to the lower Eocene in Europe and North America and to somewhat later strata in Asia. Evidently the Aplodontoidea were once widespread throughout the Palearctic and Nearctic Regions, but are now nearing the end of their evolutionary history.

The Anomaluridae are almost as restricted in central and southern Africa as the Aplodontidae in North America. They are known from Eocene and Oligocene deposits in Europe, but nowhere else. The earliest African fossils date to the lower Miocene, suggesting that they may have arisen in Europe and migrated south to persist as somewhat aberrant forms in the southern portion of Africa. Alternatively they may have arisen in Africa during the Paleocene, before the continent was isolated from Eurasia.[107]

The essential point in this brief and admittedly inadequate discussion of the Sciuromorpha is that they probably are derived from early Palearctic and Nearctic ancestors with one branch of the Sciuridae achieving considerable development in the Indo-Malayan region of the Old World Tropics. The sciurids are also distributed throughout the world and may have had an important bearing on the present distribution of ixodid ticks.

C. Myomorpha

Eurasia and Africa and, to a lesser extent, North America are now major centers of development of the suborder Myomorpha, although the fossil record clearly shows that this group arose in the Palearctic Region. There are only three superfamilies in the suborder, fewer than in either of the other suborders, but the myomorph rodents are truly world-wide in distribution. The two smaller superfamilies, Gliroidea and Dipodoidea, are presently found in Asia, Europe, and Africa. Their fossil record suggests that they developed in Europe and Asia, or both, and later spread into Africa.

Two small families of the large superfamily Muroidea, the Spalacidae and Rhizomyidae, are restricted to Europe, Africa, and Asia. Of little significance to rickettsial evolution, they are burrowing forms whose phylogeny is in some doubt, but they appear to be offshoots from the cricetid line. The Rhizomyidae is the only muroid family represented in Pleistocene deposits in Africa. The absence of pre-Pleistocene muroid fossils in Africa suggests that the Muroidea migrated there from Eurasia during recent geologic history.[118] Thus, the Palearctic must be considered the initial center of development of the Muroidea with later secondary centers in the Nearctic and Ethiopian Regions and an important center of one major group (Muridae) in Southeast Asia and contiguous island chains.

The two major families of the Muroidea, the Cricetidae and Muridae, have played an important role in the distribution and evolution of ticks and rickettsiae. Ellerman[100] considers the Cricetidae as a subfamily (Crecetinae) of the Muridae, but for the present purposes, it makes little difference how they are classified. The Cricetidae are now essentially the New World branch and the Muridae the Old World branch of the complex. Although the murids are currently more widely distributed than the cricetids, the latter are found in Oligocene deposits of both Europe and North America, as well as Asia, indicating that they were formerly more widespread than the Muridae.[99] The fossil record also suggests that the Cricetidae are a more ancient group; the Muridae are not found in strata earlier than the Pliocene. Neither group is represented in the fossil record of Africa (however, see Lavocat[119]), South America, or Australia; migration into the southern continents occurred in recent times. Today murids reach their greatest density in the region extending from Burma to New Guinea and in central and eastern Africa. There are relatively few species in India, which lies between Africa and Southeast Asia, and has had a separate origin. In Africa, most of the murids occur south of the Sahara in savanna and secondary and primary forests. Unlike Southeast Asia, the African rainforests have few species. There are 23 genera and 84 species of murids in Africa, but only one genus (*Mus*) also occurs in Tropical Asia.[118] The most logical conclusion that can be drawn is that the Muroidea differentiated in what is now the Palearctic Region of Asia and subsequently spread throughout the world.

1. Australian Rodent Fauna

The rodent fauna of Australia is composed solely of the Muroidea, more specifically the family Muridae. Since the fossil murids date only from the Pleistocene, the last epoch of the Tertiary period, their derivation requires some comment. The sparse fauna of placental mammals in Australia consists of man, who arrived there in recent times by sea; bats, which came by air; the dingo, probably introduced by early man;[120] domestic animals and rabbits introduced by modern man, and rodents. The rodent fauna consists exclusively of members of the subfamily Murinae (Family Muridae) and the Hydromyinae, a subfamily probably derived from it,[89] and the Pseudomyinae, a group of native Australian rodents of considerable variety.* The restriction of the indigenous placental mammals to one family of rodents led Ellerman[89] to postulate that the Australian segment of the Muridae evolved there independent of the Asian Murids. This, while true to some extent, is probably not a case of parallel evolution from generalized mesethmoid mammals, since the ancestral forms would have had to reach Australia either from Africa by way of Antarctica during the Paleocene[120] or from Asia in the Cretaceous. The Asian origin of Australian rodents is almost certain, but no land connected Australia and the Asian mainland after the late Mesozoic when the marsupials could have migrated to Australia.[125] Traub,[117] however, believes (as most paleontologists do) that marsupials evolved in the Southern Hemisphere and migrated back and forth between Australia and South America along island stepping stones as the continents drifted apart. Traub also questions the Asian origin of murids and postulates that they arose, or at least developed, in the Australian Region. From there they moved back and forth among the islands progressing slowly northwards until continental Asia was reached. African murids also may have been derived from Australian forms which migrated to Africa by island hopping before the continents drifted too far apart.

The ease with which organisms can move across water is well illustrated by the repopulation of islands denuded of life by natural forces. For example, Krakatau, a volcanic island in the Sunda Strait between Java and Sumatra, erupted violently in 1883, almost completely destroying all life on the island. Within 50 years the island was once again covered by dense tropical forest with more than 250 species of plants and 47 species of vertebrates as well as other animals. Most, if not all, of the new floral and faunal elements must have arrived by sea or air. In this case the distances involved were not great, Krakatau is only 25 miles from Java and 11.5 miles from Sebesi, but neither was the time span. Thus the fact of overseas dispersal of vertebrates has been established in recent times.[126] However, if Australia's placental mammals arose from generalized ancestors from either Africa or Asia, a much richer fauna should have developed. Even more important, the fossil record is devoid of mammalian forms until the late Tertiary.[120]

Towards the close of the Tertiary a fall in sea level, probably combined with earth movements, is known to have produced land emergence that culminated during the Pleistocene in the union of Sumatra, Java, and Burma with the mainland of Asia to

*The history of the Papuan-Australian rodent fauna is actually more complex than it appears at first glance. Simpson[121] gives a good, albeit brief, account of the origin of Australian rodents based primarily on the monographic works of Tate.[122,123] He divides the murids into four main groups: (1) the genus *Rattus*, (2) Old Papuan genera, mostly on New Guinea, (3) the *Pseudomys* group, for the most part confined to Australia, and (4) the subfamily Hydromyinae, mostly on New Guinea. Each of these groups probably had a different history and arrived in the Papuan-Australian Region from Southeast Asia at different times. The genus *Rattus* reached Australia most recently, first in the Pleistocene and later through introductions by man. The ancestors of the Old Papuan group reached New Guinea in the Pliocene or perhaps earlier and differentiated there. Later, the two genera, *Uromys* and *Melomys* migrated to Australia.

The Pseudomyinae, the so-called native rats of Australia, are tremendously varied as a result of adaptive radiation into different available ecological niches. They probably reached Australia from Asia via New Guinea, which was repeatedly and broadly connected to Australia for part of the Tertiary and during the Pleistocene. Their ancestors may have been the first rodents to reach Australia. The extent of their subsequent development led Simpson[121] to conclude that the ancestral forms must have been of early generalized murid stock that reached there no later than the Miocene via the Indonesian island migration route.

The Hydromyinae are primarily Papuan; only two genera reached Australia. In Simpson's opinion, the group evolved in New Guinea from a very early murid or possibly pre-murid cricetid ancestor from Asia that migrated to New Guinea through the Indonesian island chain. Readers wishing a more comprehensive treatment of this fascinating subject are referred to the excellent papers of Keast[24] and Traub.[117,124]

form Sundaland. At the same time, the shallow Sahul Shelf between New Guinea and Australia became dry land, and the connection persisted until about 8000 years ago.[127] Thus, by migration from Asia across Sundaland and island hopping eastward through Celebes, Ceram, and other islands to New Guinea, and finally across the Sahul Shelf, rodents could have reached Australia in the Pleistocene. An alternative and more likely hypothesis is that murids first reached Australia and New Guinea from Asia on more than one occasion in the late Tertiary Period by island hopping via driftwood rafts before and during the massive tectonic movements at the close of the period. Later migrations undoubtedly took place in the Quaternary Period, but these probably provided only minor additions to the already developing fauna. Placing the initial arrival of rodents to Australia in the late Tertiary (late Miocene or Pliocene) would allow a sufficient time for the fauna to develop to its present state.

2. *Rodent Fauna of the Oriental Region*

It is an unfortunate fact that the fossil record of the Oriental Region is not very complete. We can rely only on the present distribution of living murids and their affinities to those in other regions to gain some insight into its faunal history (Table 4). The area includes the Indonesian and Philippine Archipelagoes, Southeast Asia, Burma, and India south of the Himalayas. It is a vast and tremendously complex region made up of three distinct and relatively isolated areas:[117] (1) the subcontinent of India, (2) the Burmo-Chinese subregion including Assam and Indochina south to North Malaya, and (3) the Malay Peninsula extending through the Indonesian and Philippine Archipelagoes. Even within the great archipelagoes there are distinct faunal assemblages. The Philippine fauna differs in many ways from the fauna of Borneo, Java, and the mainland. Moreover, the fauna of Celebes is, in some respects, unique to that island.

a. India

The large subcontinent of India has been isolated from Africa since the Cretaceous. The Himalayas arose in the late Paleocene or early Pleistocene after the collision of India and Asia during the late Eocene and early Oligocene, imposing effective, but not completely impervious, barriers to the immigration of faunal elements from Europe, Asia, and Southeast Asia. The presence of fossil Hystricidae and Ctenodactylae suggests an early Eurasian fauna in India before this area was sealed off from the North. The presence of *Platacanthomys lasiurus* in southern India further suggests that the early Indian fauna contained East Asian representatives. The closest living relatives of *P. lasiurus, Typhlomys* spp., are now found in South China and Indochina.

In both India and Southeast Asia, the Sciuridae form a conspicuous portion of the fauna, but, whereas ground squirrels are well developed in Southeast Asia, only two genera of nonflying squirrels are found in India: *Funambulus* is confined to the subcontinent, *Ratufa* ranges eastward to Borneo. Flying squirrels are well developed in both areas.

The small muroid family, Rhizomyidae, is confined to Southeast Asia and the Eastern Himalayas of Nepal, Assam, and Burma. The distribution of the Muridae in this region also demonstrates the unique position of India compared to that of the eastern portion of the Oriental region. Of the approximately 45 genera, only 12 are known from peninsular India, whereas there are 30 in Southeast Asia. Northern India, Assam, and Burma contain 11 genera (Table 4). It is important to note that only four genera *Vandeleuria, Rattus, Mus, Bandicota* are common to both Southeast Asia and Peninsular India. *Vandeleuria* is primarily Indian with a few species in the Indochina peninsula. *Rattus* achieves its most spectacular development in Southeast Asia.[128] Only three of

Table 4
DISTRIBUTION OF GENERA OF MURIDAE IN THE ORIENTAL REGION

Genera	N.W. India	Penin. India	Assam, Burma, Nepal, Indochina	Malaysia	Philippines
Hapalomys			X		
Lenomys				X	
Chiropodomys			X[a]	X	X
Vandeleuria		X	X		
Micromys			X		
Apodemus			X		
Carpomys					X
Batomys					X
Pithecheir				X	
Ceratomys					X
Malomys				X	
Hadromys			X		
Millardia	X	X	X		
Pyromys	X				
Dacnomys			X		
Eropeplus				X[b]	
Rattus	X	X	X	X	X
Apomys					X
Coelomys		X[c]			
Haeromys				X	
Chiromyscus			X		
Mus	X	X	X	X	X
Mycteromys				X	
Crunomys					X
Golunda	X	X	X		
Echiothrix				X	
Melasmothrix				X	
Acomys	X				
Bandicota	X	X	X	X	
Nesokia	X				
Phloeomys					X
Rhynchomys					X
Celaenomys					X
Chrotomys					X
Gerbillus	X				
Tatera	X	X			
Meriones	X				
Eothenomys			X		
Anteliomys			X[a]		
Alticola			X[d]		
Microtus			X[a]		
Phaiomys			X[e]		
Neodon			X		
Limnomys					X
Tarsomys					X
Tryphomys					X
Total	11	8	19	12	15

[a]Including Yunnan Province of South China.
[b]Celebes only.
[c]Ceylon only.
[d]Kashmir.
[e]North India to Nepal.

Ellerman's subgroups *blandfordi, cutchicus* and *rattus* are found in peninsular India, the first two are found nowhere else. *Bandicota* is more evenly distributed throughout the Oriental Region, but has developed mainly in India. Two subgroups of *Mus* (*booduga* and *leggadilla*) are almost completely confined to the India portion of the Oriental Region.

Almost as many genera are common to India and Africa as to India and Southeast Asia. Only *Mus* is distributed in all three areas. Between Africa and India is a wide region of desert and semidesert populated mostly by Palearctic rodents such as the Gerbilidae and Dipodidae. *Acomys* ranges through much of Africa and the southern Palearctic, with only one species (*A. cahirimus*) coming into northwest India. *Gerbillus* has a similar distribution; in India it is essentially confined to the northwest. *Tatera* also had its greatest development in Africa, but several species in one group (*Indica*) range into Persia and down to southern India.

Much of the Indian peninsula contains Cretaceous rock beds called the Gondwana System, terrestrial in origin. In its geology and animal and plant fossils, it bears a striking resemblance to the Karroo System of South Africa, testifying to its origin as part of the supercontinent, Gondwanaland, until the great breakup, which may not have occurred until the Cretaceous, somewhat later than the breakup of South America and Africa. On the other hand, Hurley[111] suggests that East Africa and India may have begun to break apart as early as the Permian. In any event, there is little evidence that the Indian rodent fauna is derived from early African elements. During its long migration from Africa to Asia, the subcontinent was probably devoid of mammalian elements or had a unique fauna which subsequently died out without leaving any trace. More likely, the Indian fauna arose partly from early Eurasian migrants (Eocene and Oligocene and partly from later migrants from West and Southeast Asia). Since the Oligocene, India has been more or less isolated from the rest of Asia and Eurasia by the Arabian Sea to the West, the Bay of Bengal to the East, and later by the stupendous Himalayan mountain range in the North. This mighty chain, which arose after India collided with Asia, extends all the way across southern Pakistan, India, and Asam. In the West along the Afghan Border, the Safed Koh, Sulaiman, and Kirthar mountains form a southern extension running all the way to the Arabian Sea. In the East, the Arakam Yoma range, stretching to the Bay of Bengal, forms a barrier between Burma and India. The entire eastern portion of Burma is composed of the Shan massif extending southward into Tenasserim and forming the Indo-Malayan system. Thus, the introduction of species from East, West, or North has long been restricted by geographical barriers.

b. Philippines

The muroid fauna of the Philippines, although certainly of Asian origin, is now unique, suggesting a long period of isolation from the mainland and the Indonesian island chain. Of the 15 genera of murids known from the Philippines, only 3 (*Chiropodomys, Rattus,* and *Mus*) are also found in the rest of the Oriental Region (Table 4).

c. Southeast Asia

Southeast Asia is the center of origin and distribution of the genus *Rattus,* an event of the most profound significance to rickettsial evolution and about which we will have much more later. The area is also rich in other murid forms, containing a large fauna of sciurids. The affinities of these groups are almost entirely Asian.

It is abundantly clear that, at least in terms of rodent fauna, the Oriental Faunal Region cannot be treated as a single homogeneous system.

SUMMARY

When the rodents began to differentiate from their insectivore or insectivore-like ancestors, the land masses of the world only vaguely resembled their present distribution as isolated continents. The climatic zones were also probably extremely different, with the tropics much more extensive than at present. The basic assumption that the Rodentia arose in the Old World Tropics seems reasonable enough if it is understood that the term refers to the huge land masses emcompassing much of what is now Eurasia and Africa. Great geologic events taking place over vast periods and accompanying climatic changes and isolation of faunal elements provided the circumstances for the divergence and differentiation of the fauna. Some present-day geologic structures afford a brief glimpse of the processes involved and possibly indicate major centers of mammalian development. One such region is the Great Rift Valley of Africa extending from Mozambique to the Gulf of Aden where it splits, one branch swinging northwesterly to form what is now the Red Sea and the other disappearing in the mid-Oceanic Ridge south of the vast Indus cone in the Arabian Basin.[129] In their present forms, the great Rift Valley systems in Africa are primarily mid-Pleistocene in origin,[130] but they are situated on earlier rift structures of the Cretaceous. The mid-Atlantic Ridge and Rift Valley between Africa and South America may be the remnants of a huge archipelago resulting from the breakup of Gondwanaland during the second half of the Mesozoic.

The complex Indonesian island chain, if viewed from the perspective of geologic time, is a dynamic, ever-changing formation of small and large land masses variously separated from each other and the Asian mainland by waterways and again joined together in different ways. Such an active, changing, physical environment undoubtedly had profound effects on mammalian evolution, providing long periods of virtual isolation of faunal elements followed by periods of intense migration, mixing, and competition between groups.

The emergence of great mountain chains and vast deserts on more or less stabilized continental regions during the Tertiary period provided barriers to the interchange of faunal elements, allowing for differentiation and divergence of major mammalian groups. Obvious examples are the Himalayas separating temperate from tropical Asia and the Sahara Desert lying between temperate Europe and tropical Africa. In the New World the Andes, Sierra Nevada, and Rocky Mountains run roughly north and south along the western portions of North and South America and thus did not prevent migration of animals from Asia into North America and eventually all the way to Patagonia. The North American mountain chains, however, did more or less isolate the western from the eastern fauna of the continent.

Geographical isolation of faunal elements was never permanent and often incomplete. Thus major migration routes existed at various times and for various periods between North and South America, North America and Asia, Southeast Asia and Northeast Asia, Southeast Asia through the Indonesian chain to Australia, between Eurasia and Africa along the Great Rift Valley, between Asia and Europe, and perhaps even between Africa and South America by rafting or island hopping.

The picture I wish to paint is that of a dynamic earth, ever-changing physically as well as biologically, with these developments taking place over almost inconceivably long spans of time. Such a conception is essential to understand the processes of evolution.

In Volume I, Chapter 3 I suggested that three major Ixodid genera (*Dermacentor, Hyalomma,* and *Haemaphysalis*) originated in the Palearctic, Ethiopian, and Oriental Regions, respectively (Volume I, Chapter 3, Figure 1). These centers of origin correlate closely with those of the major families of sciuromorph and myomorph rodents (Figure

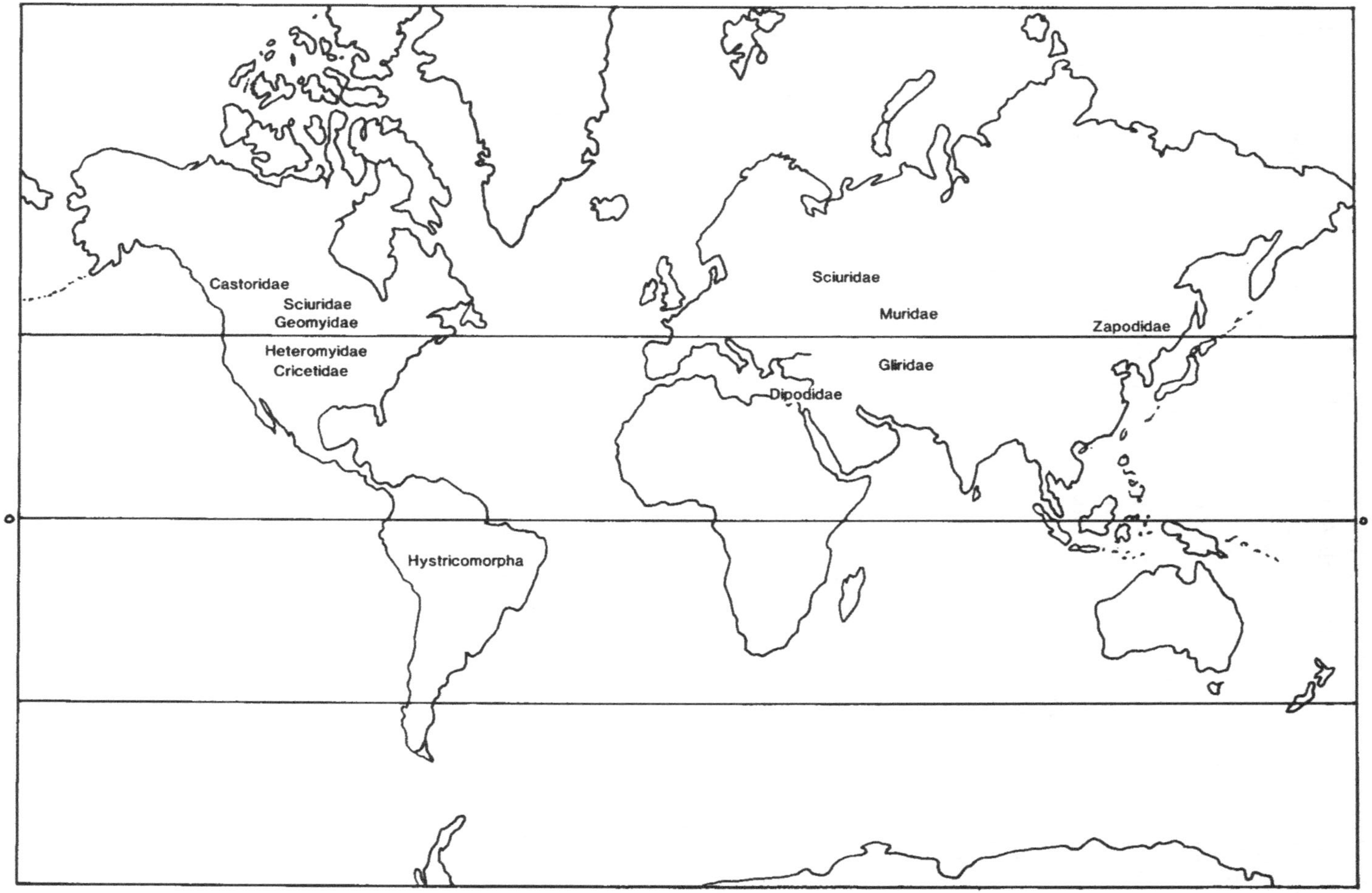

FIGURE 1. Centers of origin of major rodent groups.

1). The center of origin of *Amblyomma* in South America (or Gondwanaland) coincides with the center of evolution of the hystricomorph rodents. Three of the four small rodent families thought to have originated in Africa are hystricomorphs. This also correlates well with the presence of a large *Amblyomma* population in Africa today.

Ixodes, which now claims North America and Africa as its "home", probably developed in Laurasia and, later, became more or less isolated in the two widely separated continents. It may or may not have developed primarily as a rodent parasite. If it did so in Africa, the native hystricomorph rodents may have been its original hosts, but subsequent development must have been on other native animals as they evolved, or on immigrants from the North and East.

Rhipicephalus clearly originated in Africa, but probably did not do so as a parasite of the small group of hystricomorph rodents or the Pedetidae, which are thought to have evolved there. *Rhipicephalus* probably evolved more recently than the other genera, perhaps not until the mammalian fauna of Africa was well developed. Kaiser and Hoogstraal[131] consider it to be a still rapidly differentiating genus of "recent" origin.

Another large and extremely diverse order of mammals, the Artiodactyla, undoubtedly played a significant role in the development, dispersal, and divergence of ixodid ticks. Pilgrim[132] presents an excellent review of the evolution and zoogeography of this great mammal group. (See also Webb[133,134] for a history of savanna vertebrates in the New World.) The evolution, divergence, and dispersion of both artiodactyls and rodents were similar in many ways. The complex life histories of many ticks, often including parasitism of both rodents and herbivores by different stages of the same species, undoubtedly developed early. The contributions of the artiodactyls to ixodid development in the Eocene and later may have equalled that of the rodents, but space limitations do not allow detailed discussions of the Artiodactyla.

The correlation of the centers of origin of ixodid genera and those of various rodent families may or may not be of great evolutionary significance. At best, it is a gross oversimplification of the extremely complex forces in operation over geologic time. Nonetheless, it can form the basis of an hypothesis which, although not provable, is logical and reasonable from a phylogenetic standpoint. Conceivably mammalogists and acarologists, working together, could develop this hypothesis of tick-rodent-herbivore association as an evolutionary system. Such a collaborative study would be of great value.

With this background on the evolutionary trends and zoogeography of ixodid ticks and their major vertebrate hosts, we are now ready to consider the evolution of the microbial parasites (or symbiotes) themselves.

REFERENCES

1. **Bambach, R. K., Scotese, C. R., and Ziegler, A. M.,** Before Pangea: the geographies of the Paleozoic World, *Am. Sci.*, 68, 26, 1980.
2. **Dietz, R. S. and Holden, J. C.,** Reconstruction of Pangaea: breakup and dispersion of continents, Permian to Present, *J. Geophys. Res.*, 75, 4939, 1970.
3. **Dietz, R. S. and Holden, J. C.,** The breakup of Pangaea, *Sci. Am.*, 223, 30, 1970.
4. **Taylor, F. B.,** Bearing of the Tertiary mountain belt on the origin of the earth's plan, *Bull. Geol. Soc. Am.*, 21, 179, 1910.
5. **Wegener, A. L.,** *Die Entstehung der Kontinent und Ozeane*, 3rd ed., F. Vieweg and Son, Braunsweig, 1922, 1 (Transl. Kerl, J.G.A., *The Origin of Continents and Oceans*, Methuen and Co., London, 1924, 1.)

6. **Hess, H. H.,** History of ocean basins, *Petrologic Studies; A Volume in Honor of A. F. Buddington,* James, H. L. and Leonard, B. L., Eds., Geological Soc. Amer., New York, 1962, 1.
7. **Dietz, R. S.,** Continent and ocean basin evolution by spreading of the sea floor, *Nature (London),* 190, 854, 1961.
8. **Isacks, B., Oliver, J., and Sykes, L.,** Seismology and the new global tectonics, *J. Geophys. Res.,* 73, 5855, 1968.
9. **Bird, J. M. and Isaacs, B.,** Eds. *Plate Tectonics,* American Geophysical Union, Washington, D.C., 1972, 1.
10. **Dewey, J. F.,** Plate tectonics, *Sci. Am.* 226, 56, 1972.
11. **Dickinson, W. R.,** Plate tectonics in geological history, *Science,* 174, 107, 1971.
12. **Dickinson, W. R.,** Evidence for plate tectonics regimes in the rock record, *Am. J. Sci.,* 272, 551, 1972.
13. **McKenzie, D. P.,** Plate tectonics and continental drift, *Endeavour,* 29, 39, 1970.
14. **McKenzie, D. P.,** Plate tectonics and sea floor spreading, *Am. Sci.,* 60, 425, 1972.
15. **Tarling, D. H. and Tarling, M.,** *Continental Drift: A Study of the Earth's Moving Surface,* Doubleday, Garden City, N.Y., 1971, 1.
16. **McElhinny, M. W.,** *Paleomagnetism and Plate Tectonics,* Cambridge University Press, London, 1973. 1.
17. **Vine, F. J. and Mathews, D. H.,** Magnetic anomalies over oceanic ridges, *Nature (London),* 199, 947, 1963.
18. **Vine, F. J.,** Spreading of the ocean floor: new evidence, *Science,* 154, 1405, 1966.
19. **McKenzie, D. P. and Parker, R. L.,** The North Pacific: an example of tectonics on a sphere, *Nature (London),* 216, 1276, 1967.
20. **Morgan, W. J.,** Rises, trenches, great faults and crustal blocks, *J. Geophys. Res.,* 73, 1959, 1968.
21. **Le Pichon, X.,** Sea floor spreading and continental drift, *J. Geophys. Res.,* 73, 3661, 1968.
22. **du Toit, A. L.,** *Our Wandering Continents,* Oliver and Boyd, Edinburgh, 1937, 1.
23. **Keast, A.,** Evolution of mammals on southern continents. I. Introduction: the southern continents as backgrounds for mammalian evolution, *Q. Rev. Biol.,* 43, 225, 1968.
24. **Keast, A.,** Evolution of mammals on southern continents. IV. Australian mammals: zoogeography and evolution, *Q. Rev. Biol.,* 43, 373, 1968.
25. **Keast, A.,** Continental drift and the evolution of the biota on southern continents, *Q. Rev. Biol.,* 46, 335, 1971.
26. **Myers, G. S.,** Zoogeographical evidence of the age of the South Atlantic Ocean, *Stud. Trop. Oceanogr.,* 5, 614, 1967.
27. **Darlington, P. J., Jr.,** *Biogeography of the Southern End of the World,* Harvard University Press, Cambridge, Mass., 1965, 1.
28. **Edmunds, G. F., Jr.,** Biogeography and evolution of Ephemeroptera, *Annu. Rev. Entomol.,* 17, 21, 1972.
29. **Evans, J. W.,** A zoogeography of some Australian insects, *Monogr. Biol.,* 8, 150, 1959.
30. **Gressit, J. L.,** Subantarctic entomology and biogeography, *Pac. Insects Monogr.,* 23, 295, 1970.
31. **Gressitt, J. L.,** Antarctic entomology with emphasis on biogeographical aspects, *Pac. Insects Monogr.,* 25, 167, 1971.
32. **Colbert, E. H.,** Tetrapods and continents, *Q. Rev. Biol.,* 56, 250, 1971.
33. **Colbert, E. H.,** *Lystrosaurus* and Gondwanaland, *Evol. Biol.,* 6, 157, 1972.
34. **Colbert, E. H.,** Antarctic fossils and the reconstruction of Gondwanaland, *Nat. Hist.,* January 1972, 67.
35. **Elliott, D. H., Colbert, E. H., Breed, W., Jensen, J. A., and Powell, J. S.,** Triassic tetrapods from Antarctica: evidence for continental drift, *Science,* 169, 1197, 1970.
36. **Kitching, J. W., Collinson, J. W., Elliot, D. H., and Colbert, E. H.,** *Lystrosaurus* Zone (Triassic) fauna from Antarctica, *Science,* 175, 524, 1972.
37. **Cracraft, J.,** Continental drift and vertebrate distribution, *Annu. Rev. Ecol. Syst.,* 5, 215, 1974.
38. **Cox, C. B.,** Migrating marsupials and drifting continents, *Nature (London),* 226, 767, 1970.
39. **Fooden, J.,** Breakup of Pangaea and isolation of relict mammals in Australia, South America and Madagascar, *Science,* 175, 894, 1972.
40. **Martin, P. G.,** The Darwin Rise hypothesis of the biogeographical dispersal of marsupials, *Nature (London),* 225, 197, 1970.
41. **Lillegraven, J. A.,** Biogeographical considerations of the marsupial-placental dichotomy, *Annu. Rev. Ecol. Syst.,* 5, 263, 1974.
42. **Schlinger, E. I.,** Continental drift, *Nothofagus* and some ecologically associated insects, *Annu. Rev. Ecol. Syst.,* 5, 323, 1974.

43. **Raven, P. H. and Axelrod, D. I.,** Plate tectonics and Australasian paleobiogeography, *Science,* 176, 1379, 1972.
44. **Bullard, E., Everett, J. E., and Smith, A. G.,** The fit of the continents around the Atlantic, *Philos. Trans. R. Soc. London Ser. A,* 258, 41, 1965.
45. **LePichon, X. and Fox, P. J.,** Marginal offsets, fracture zones and the early opening of the Atlantic, *J. Geophys. Res.,* 76, 6294, 1971.
46. **Pitman, W. C., III and Talwani, M.,** Sea floor spreading in the North Atlantic, *Bull. Geol. Soc. Am.,* 83, 619, 1972.
47. **Dietz, R. S. and Sproll, W. P.,** Fit between Africa and Antarctica: a continental drift reconstruction, *Science,* 167, 1612, 1970.
48. **Griffiths, J. R.,** Reconstruction of the Southwest Pacific margin of Gondwanaland, *Nature (London),* 234, 203, 1971.
49. **Meservey, R.,** The coastline fit of Africa and South America, *Palaeogeogr. Palaeoclimatol. Palaeoecol.,* 9, 233, 1971.
50. **Smith, A. G. and Hallam, A.,** The fit of the southern continents, *Nature (London),* 225, 139, 1970.
51. **Sproll, W. P. and Dietz, R. S.,** Morphological continental drift fit of Australia and Antarctica, *Nature (London),* 222, 345, 1969.
52. **King, L. C.,** Necessity for continental drift, *Am. Assoc. Petrol. Geol. Bull.,* 37, 2163, 1953.
53. **Adie, R. J.,** Geological evidence on possible antarctic land connections, in *Pacific Basin Biogeography,* Gressitt, J. L., Ed., 10th Pac. Sci. Congr., Honolulu, 1963, 455.
54. **Allard, G. O. and Hurst, V. J.,** Brazilian-Gabon geologic links supports continental drift, *Science,* 165, 528, 1969.
55. **Craddock, C.,** Antarctic geology and Gondwanaland, *Antarct. J.,* 5, 53, 1970.
56. **Hallam, A.,** Mesozoic geology and the opening of the North Atlantic, *J. Geol.,* 79, 129, 1971.
57. **Hurley, P. M. and Rand, J. R.,** Pre-drift continental nuclei, *Science,* 164, 1229, 1969.
58. **Fraker, L. A. and Crowell, J. C.,** Geological evidence for the place of Antarctica in Gondwanaland, *Antarct. J.,* 5, 67, 1970.
59. **Barker, P. F.,** Plate tectonics of the Scotia Sea, *Nature (London),* 228, 1293, 1970.
60. **Creer, K. M., Embleton, B. J. J., and Valencio, D. A.,** Triassic and Permo-Triassic paleomagnetic data for South America, *Earth Planet. Sci. Lett.,* 8, 173, 1970.
61. **McElhinny, M. W. and Luck, G. R.,** Paleomagnetism and Gondwanaland, *Science,* 168, 830, 1970.
62. **Vilas, J. F. and Valencio, D. A.,** Paleographic reconstructions of the Gondwana continents based on paleomagnetic and sea floor spreading data, *Earth Planet. Sci. Lett.,* 7, 397, 1970.
63. **Weissel, J. K. and Hayes, D. E.,** Asymmetric sea floor spreading south of Australia, *Nature (London),* 231, 518, 1971.
64. **Briggs, P.,** *200,000,000 Years Beneath the Sea,* Holt, Reinhart & Winston, New York, 1971, 1.
65. **Calder, N.,** *The Restless Earth, A Report on the New Geology,* Viking, New York, 1972, 1.
66. **Heezen, B. C. and Hollister, C. D.,** *The Face of the Deep,* Oxford University Press, London, 1971, 1.
67. **Hughes, N. F., Ed.,** Organisms and continents through time: a symposium, *Spec. Pap. Palaeontol.,* 12, 1, 1973.
68. **Scientific American Book,** *The Ocean,* W. H. Freeman, San Francisco, 1969, 1.
69. **Scientific American Book,** *Continents Adrift,* W. H. Freeman, San Francisco, 1971, 1.
70. **McKenzie, D. P. and Sclater, J. G.,** The evolution of the Indian Ocean, *Sci. Am.,* 228, 63, 1973.
71. **Markl, R. G.,** Evidence for the breakup of eastern Gondwanaland by the early Cretaceous, *Nature (London),* 251, 196, 1974.
72. **Johnson, B. D., Powell, C. McA., and Veevers, J. J.,** Spreading history of the eastern Indian Ocean and greater India's northward flight from Antarctica and Australia, *Bull. Geol. Soc. Am.,* 87, 1560, 1976.
73. **Larson, R. L. and Ladd, J. W.,** Evidence for the opening of the South Atlantic in the early Cretaceous, *Nature (London),* 246, 209, 1973.
74. **Le Pichon, X. and Hayes, D. E.,** Marginal offsets, fracture zones and the early opening of the South Atlantic, *J. Geophys. Res.,* 76, 6283, 1971.
75. **Maxwell, A. E., Von Herzen, R. P., Jinghwa Hsu, K., Andrews, J. E., Saito, T., Percival, S. F., Jr., Milow, E. D., and Boyce, R. E.,** Deep sea drilling in the South Atlantic, *Science,* 168, 1047, 1970.
76. **Jones, J. G.,** Australia's Cenozoic drift, *Nature (London),* 230, 237, 1971.
77. **Le Pichon, X. and Heirtzler, J. R.,** Magnetic anomalies in the Indian Ocean and sea floor spreading, *J. Geophys. Res.,* 73, 2101, 1968.
78. **Weissel, J. K. and Hayes, D. E.,** Magnetic anomalies in the southeast Indian Ocean, *Antarct. Res. Ser.,* 19, 165, 1972.

79. **Fooden, J.**, Rifting and drift of Australia and the migration of mammals, *Science*, 180, 759, 1973.
80. **Whitmore, F. C., Jr. and Stewart, R. H.**, Miocene mammals and Central American seaways, *Science*, 148, 180, 1965.
81. **Klootwijk, C. T. and Pierce, J. W.**, India's and Australia's pole path since the late Mesozoic and the India-Asia collision, *Nature (London)*, 282, 605, 1979.
82. **Flores, G.**, Suggested origin of the Mozambique channel, *Trans. Geol. Soc. S. Afr.*, 73, 1, 1970.
83. **Darlington, P. J., Jr.**, Area, climate and evolution, *Evolution*, 13, 488, 1959.
84. **Durham, J. W.**, Paleogeographic conclusions in light of biological data, in *Pacific Basin Biogeography-Tropics*, Gressitt, J. L., Ed., Bishop Museum Press, Honolulu, 1963, 355.
85. **Durham, J. W.**, Paleoclimates, *Phys. Chem. Earth*, 3, 1, 1959.
86. **Schwarzbach, M.**, *Climates of the Past*, (transl. by Richard O. Muir), D Van Nostrand, London, 1963, 1.
87. **Dobzhansky, T.**, Evolution in the Tropics, *Am. Sci.*, 38, 209, 1950.
88. **Wood, A. E.**, Eocene radiation and phylogeny of the rodents, *Evolution*, 13, 354, 1959.
89. **Ellerman, J. R.**, *The Families and Genera of Living Rodents*, Vol. 1, British Museum (Natural History), London, 1940, 1.
90. **Simpson, G. G.**, History of the fauna of Latin America, *Am. Sci.*, 38, 361, 1950.
91. **Mayr, E.**, History of the North American Bird fauna, *Wilson Bull.*, 58, 1, 1946.
92. **Simpson, G. G.**, *Evolution and Geography*, University of Oregon Press, Salem, 1953, 1.
93. **Wilson, R. W.**, Evolution of the early Tertiary rodents, *Evolution*, 5, 207, 1951.
94. **Simpson, G. G.**, Holarctic mammalian faunas and continental relationships during the Cenozoic, *Bull. Geol. Soc. Am.*, 58, 613, 1947.
95. **Russell, L. S.**, Fossil mammals and intercontinental connections in evolution, in *Evolution: Its Science and Doctrine*, Cameron, T. W., Ed., University of Toronto Press, Toronto, 1960, 63.
96. **McKenna, M. C.**, Fossil mammals and early Eocene North Atlantic land continuity, *Ann. M. Bot. Gard.*, 6, 335, 1975.
97. **Cox, C. B.**, Systematics and plate tectonics in the spread of marsupials, *Spec. Pap. Paleontol.*, 12, 113, 1973.
98. **Tedford, R. H.**, Marsupials and the new paleogeography, *Soc. Econ. Paleontol. Mineral. Spec. Pub.*, 21, 109, 1974.
99. **Wood, A. E.**, Rodents—a study in evolution, *Evolution*, 1, 154, 1947.
100. **Ellerman, J. R.**, *The Families and Genera of Living Rodents*, Vol. 2, British Museum (Natural History), London, 1941. 1.
101. **Ellerman, J. R.**, *The Families and Genera of Living Rodents*, Vol. 3, Part I, British Museum (Natural History), London, 1949, 1.
102. **Simpson, G. G.**, The principles of classification and a classification of mammals, *Bull. Am. Mus. Nat. Hist.*, 85, 1, 1945.
103. **Patterson, B. and Pascual, R.**, Evolution of mammals on southern continents. V. The fossil mammalian fauna of South America, *Q. Rev. Biol.*, 43, 409, 1968.
104. **Hershkovitz, P.**, Evolution of mammals on southern continents. VI. The recent mammals of the Neotropical region: a zoogeographic and ecological review, *Q. Rev. Biol.*, 44, 1, 1969.
105. **Orlosky, F. J. and Swindler, D. R.**, Origins of the New World monkeys, *J. Human Evol.*, 4, 77, 1975.
106. **Gidley, J. W.**, Paleocene primates of the Fort Union, with discussion of relationships of Eocene primates, *Proc. U.S. Natl. Mus.*, 63, 1, 1923.
107. **Cooke, H. B. S.**, Evolution of mammals on southern continents. II. The fossil mammal fauna of Africa, *Q. Rev. Biol.*, 43, 234, 1968.
108. **Fairbridge, R. W.**, The Indian Ocean and the status of Gondwanaland, *Prog. Oceanogr.*, 3, 83, 1965.
109. **Creer, K. M.**, Palaeomagnetic data and du Toit's reconstruction of Gondwanaland, *Nature (London)*, 204, 369, 1964.
110. **Fell, H. B.**, Cretaceous and Tertiary surface currents of the oceans, *Annu. Rev. Oceanogr. Mar. Biol.*, 5, 317, 1967.
111. **Hurley, P. M.**, The confirmation of continental drift, *Sci. Am.*, 218, 53, 1968.
112. **Hoffsetter, R.**, Relationships, origins, and history of the ceboid monkeys and cavimorph rodents: a modern reinterpretation, *Evol. Biol.*, 6, 323, 1972.
113. **Smith, A. G. and Briden, J. C.**, *Mesozoic and Cenozoic Paleocontinental Maps*, Cambridge University Press, Cambridge, England, 1977, 1.
114. **Gingerich, P. D.**, Evolutionary patterns in early Cenozic mammals, *Annu. Rev. Earth Planet Sci.*, 8, 407, 1980.
115. **Black, C. C.**, Holarctic evolution and dispersal of squirrels. (Rodentia: Sciuridae), *Evol. Biol.*, 6, 305, 1972.

116. **Mein, P.**, Les Sciuropteres (Mammalia, Rodentia), Neogenes d'Europe occidentale, *Geobios (Lyon)*, 3, 7, 1970.
117. **Traub, R.**, Notes on zoogeography, convergent evolution and taxonomy of fleas (Siphonaptera), based on collections from Gunong Benom and elsewhere in Southeast Asia. III. Zoogeography, *Bull. Br. Mus. Nat. Hist. Zool.*, 23, 389, 1972.
118. **Missone, X.** African and Indo-Australian Muridae. Evolutionary trends, *Ann. Mus. R. Afr. Cent. Tervuren Belg., Ser. 8° Sci. Zool.*, 172, 1, 1969.
119. **Lavocat, R.**, Origine et affinites des rongeurs de sous-famille de Dendromurinae, *C. R. Acad. Sci.*, 248, 1353, 1959.
120. **Cameron, T. W. M.**, Southern intercontinental connections and the origin of the southern mammals, in *Evolution: Its Science and Doctrine*, Cameron, T. W. M., Ed., University of Toronto Press, Toronto, 1960, 79.
121. **Simpson, G. G.**, Historical zoogeography of Australian mammals, *Evolution*, 15, 431, 1961.
122. **Tate, G. H. H.**, Some Muridae of the Indo-Australian Region, *Bull. Am. Mus. Nat. Hist.*, 72, 501, 1936.
123. **Tate, G. H. H.**, The rodents of Australia and New Guinea, *Bull. Am. Mus. Nat. Hist.*, 97, 183, 1951.
124. **Traub, R.**, The zoogeography of fleas (Siphonaptera) as supporting the theory of continental drift, *J. Med. Entomol.*, 9, 584, 1972.
125. **Gill, E. D.**, The paleogeography of Australia in relation to the migration of marsupials and men, *Trans. N.Y. Acad. Sci.*, 28, 5, 1965.
126. **Thorne, R. F.**, Biotic distribution patterns in the tropical Pacific, in *Pacific Basin Biogeography*, Gressitt, J. L., Ed., Bishop Museum Press, Honolulu, 1963, 311.
127. **Jennings, J. N.**, Some attributes of Torres Strait, in *Bridge and Barrier*, Walker, D., Ed., Australian National University, Canberra, 1972, 29.
128. **Schwarz, E. and Schwarz, H. K.**, A monograph of the *Rattus rattus* group, *An. Esc. Nac. Cienc. Biol. Mexico City*, 4, 79, 1965 (1967).
129. **National Geographic Society**, Map of the Indian Ocean Floor, *Natl. Geogr.*, October 1967.
130. **du Toit**, A. L., *Geology of South Africa*, 3rd ed., Haughton, S. H., Ed., Hafner Publishing, New York, 1954, 1.
131. **Kaiser, M. N. and Hoogstraal, H.**, Noteworthy recent tick records from Egypt. II. Hosts, distribution and ecology of *Rhipicephalus simus* Koch, *J. Egypt. Public Health Assoc.*, 42, 231, 1967.
132. **Pilgrim, G. E.**, The dispersal of the Artiodactyla, *Biol. Rev.*, 16, 134, 1941.
133. **Webb, S. D.**, A history of savanna vertebrates in the New World. I. North America, *Annu. Rev. Ecol. Syst.*, 8, 355, 1977.
134. **Webb, S. D.**, A history of savanna vertebrates in the New World. II. South America and the great interchange, *Annu. Rev. Ecol. Syst.*, 9, 393, 1978.

Chapter 5

THE TICKBORNE RICKETTSIAE OF THE SPOTTED FEVER OR TICK TYPHUS GROUP

N. Marchette

TABLE OF CONTENTS

I. Characteristics of the Group 76
- A. Antigenic Characteristics 76
- B. Classification of Tickborne Rickettsiae 77

II. Geographical and Host Distribution 78
- A. *Rickettsia conorii* 78
  - 1. Natural Cycles in Africa 78
  - 2. *R. conorii* in India 80
  - 3. Tick-Typhus Strains in Southeast Asia 80
  - 4. Dispersion from Center of Origin 81
- B. *Rickettsia sibirica* 82
  - 1. Southern Limits 82
  - 2. Western Limits 84
  - 3. Southeastern Limits 85
- C. *Rickettsia rickettsii* 85
  - 1. Acarine Hosts 85
    - a. *Dermacentor andersoni* 85
    - b. *Dermacentor variabilis* 87
    - c. *Amblyomma cajennense* 87
    - d. *Amblyomma americanum* 87
    - e. Other Amblyommids 88
    - f. *Haemaphysalis leporispalustris* 88
  - 2. Wild Vertebrate Hosts 88
    - a. The Role of Lagomorphs in *R. Rickettsii* Ecology and Evolution 88
    - b. Origin and Dispersion 91
- D. *Rickettsia parkeri* 92
- E. *Rickettsia montana* 92
- F. *Rickettsia rhipicephali* 95
- G. *Rickettsia akari* 96
  - 1. Rickettsial pox 96
  - 2. Distribution 96
  - 3. Natural Hosts 96
  - 4. Affinities 97
  - 5. Summary 97
- H. *Rickettsia australis* 97
  - 1. Natural Hosts 98
  - 2. Affinities 98
  - 3. Origin 98

I. Newly Described Tick-Typhus Rickettsiae 99
1. *Ixodes pacificus* Strain 99
2. *Dermacentor parumapertus* Strain 99
3. *Dermacentor occidentalis* Strain 99
4. *Hlp*-like *R. rickettsii* Strains 99
5. *Amblyomma americanum* Strain 100

III. Evolution of the *Rickettsia (Dermocentroxenus)* Tickborne Line 100

References 103

I. CHARACTERISTICS OF THE GROUP

Several groups of organisms currently classified in the Order Rickettsiales are transmitted by ticks or are more or less closely associated with ticks. *Wolbachia persica* in the Tribe Wolbachieae in the Family Rickettsiaceae is a symbiote of argasid ticks, and many, if not all, ticks contain rickettsia-like (or *Wolbachia*-like) symbiotes which are as yet unnamed and unclassified. *Wolbachia* and other genera in the Tribe Wolbachieae include symbiotes of a variety of arthropods, insects as well as acarines (see Volume II, Chapter 8). Some members of the family Anaplasmataceae are also tickborne or ecologically associated with ticks (Volume II, Chapter 11).

Thus, the term "tickborne rickettsiae" could refer to a variety of related and unrelated organisms if it is not further qualified. Tick typhus and spotted fever are terms describing the illness caused by a subgroup of the genus *Rickettsia*. Philip[1] placed these organisms in the subgenus *Dermacentroxenus* to distinguish them from the insectborne *Rickettsia* and the trombiculid miteborne *Zinssera*. In the latest *Bergey's Manual* Moulder[2] does not include these terms because he doesn't believe in subgenera. This may be a reasonable attitude, but *Dermacentroxenus* is a useful term to designate those tick symbiotes which are transmissible to vertebrates and possess common group antigens detectable by serological techniques. Spotted fever and, to a lesser extent, tick typhus are widely used in the literature to refer to the organisms of the *Dermacentroxenus* line as well as, more correctly, to the disease they cause in man. In the following discussion, spotted fever group and tick typhus group rickettsiae are synonymous with *Dermacentroxenus*.

A. Antigenic Characteristics

The spotted fever group rickettsiae were antigenically defined by Lackman and Pickens[3] on the basis of a common soluble antigen that fixed complement in the presence of convalescent Rocky Mountain spotted fever guinea pig serum. Within this group are a number of organisms possessing unique antigenic determinants variously detectable by cross-infection and vaccine-protection tests,[4–12] complement fixation tests with washed CF antigens,[8,12–16] complement fixation tests with species-specific antibodies produced in mice,[4,17] the rickettsial toxin neutralization test,[18,19] and a microimmunofluorescence test.[20] On the basis of these tests, particularly the microimmunofluoresence test, as adapted by Philip et al.,[20] immunological relationships among the tick typhus serotypes have been determined.

Table 1
THE TICKBORNE (SPOTTED FEVER GROUP) RICKETTSIAE AND THEIR DISTRIBUTION

Rickettsia conorii	Africa, Middle East, southern Europe, South Asia
R. sibirica	Asia, Eurasia
R. rickettsii	North and South America
R. slovaca	Eastern Europe
R. akari	Eastern Asia, Eastern North America
R. australis	Australia
R. parkeri	North America
R. canada	North America
R. montana	North America
R. rhipicephali	North America
Ixodes pacificus strain	North America
Dermacentor occidentalis strain	North America
D. parumapertus strain	North America
D. variabilis strain	North America
Haemaphysalis leporispalustris strain	North America
Amblyomma americanum strain	North America
JC 880	Pakistan
Thai TT-118	Thailand
Swiss agent	Switzerland

B. Classification of Tickborne Rickettsiae

The present official classification of tickborne rickettsiae includes: *R. conorii,* principally in Africa, Europe, and South Asia; *R. sibirica* in Asia proper; *R. rickettsii, R. parkeri,* and *R. canada* in the New World; *R. australis* and *R. akari* in Asia and North America.[2] Two other tickborne rickettsiae in the U.S., *R. montana* and *R. rhipicephali,* and one in Europe, *R. slovaca,* have been described and named since publication of the 8th edition of *Bergey's Manual*. A number of new serotypes also have been isolated, but not yet named (Table 1). Rickettsial strains recovered from *Ixodes pacificus* in Oregon, *Dermacentor occidentalis* in California, and *D. parumapertus* in Nevada and Utah appear to be antigenically distinct from other tick typhus rickettsiae and are considered to be serotypes of the spotted fever group. Single strains of three rickettsiae from *Haemaphysalis leporispalustris* in Canada, *Amblyomma americanum* in Alabama, and *D. variabilis* in Arkansas, respectively, differ antigenically from all other rickettsiae, and their relationships are not known.[20] Additional serologically distinct spotted fever group rickettsiae will probably be discovered in the near future.[21]

The complexity of the tick typhus group in North America is apparent, and there is evidence that a similar variety of serotypes exists in other parts of the world. In the Indian subcontinent, strains of *R. conorii* and *R. sibirica,* as well as several isolates of a distinct serotype have been reported;[19,22] and a strain isolated in Thailand differs antigenically from all other tick typhus serotypes.[19]

In Eastern Europe, strains distinct from *R. sibirica* also have been isolated, and strains from ticks in Czechoslovakia[23] have been characterized as serologically distinct from prototype spotted fever group species and named *R. slovaca.*[24] Antibody to *R. slovaca* is prevalent among hares (*Lepus europaeus*) in Central Slovakia,[25,26] but they develop only inapparent infections when inoculated with the agent.[27] Strains isolated from wild rodents are reported to be identical to *R. sibirica,*[28] and an analysis of strains isolated in the Armenian S.S.R., some from *Argas persicas* ticks,[29] indicate that they can all be included in the *R. sibirica* serotype.[30] It should be noted that serological analysis of strains is not without its difficulties and results from one laboratory may not be exactly comparable to those from another. Vertebrate species differ considerably in their ability to produce antibody to foreign antigens; and, depending on which species or strain is used, an organism may be classified as antigenically distinct or identical to one or more recognized agents.[31]

Still unnamed is the "Swiss agent," a spotted fever group rickettsia naturally infecting *I. ricinus* in Switzerland.[32] Unlike most strains of *R. rickettsii*, it is not pathogenic for guinea pigs, rabbits, or mice, but produces microscopic lesions in the tunica vaginalis of male voles. The relationship of this organism to other spotted fever species is unclear at this time.

The habitats of these rickettsiae are so diverse, ranging from deserts to tropical rain forest to temperate deciduous forests to alpine formations, and their host range so wide that antigenic heterogeneity is not surprising. More intensive investigation of the organisms indigenous to other major biomes and faunal zones (Africa, North and Central Asia, South America) would almost surely uncover other distinct rickettsial species or, at least, distinct serotypes.

II. GEOGRAPHICAL AND HOST DISTRIBUTION

A. *Rickettsia conorii*

Rickettsia conorii is enzootic in southern Europe, Israel, Turkey, the Crimea (Black Sea Basin), the Caspian Sea Basin, India, West Pakistan, Kashmir, Burma, and possibly Southeast Asia. It is widespread throughout Africa,[33] but has been studied most intensively in South Africa,* Kenya, and the countries bordering the Mediterranean.

1. Natural Cycles in Africa

There are hundreds of accounts of *fievre boutonneuse* in man in Southern Europe and Africa and its relationship to the common dog tick, *Rhipicephalus sanguineus*, yet there is little information about the natural cycles that must exist in the wild rodent and tick fauna. *Rickettsia conorii* has been reported in at least 25 species of ticks (Table 2), but the mere isolation of a rickettsia from a tick in no way establishes the vector potential of the arthropod or its basic importance in natural cycles. Experimental transmission has been demonstrated in only four species: *Haemaphysalis leachii, Ixodes ricinus, Rhipicephalus appendiculatus*, and *R. sanguineus*. Gear's opinion[44] that most ixodid ticks of Africa can transmit the agent may be exaggerated, but it is probably not too wide of the mark. The "home" of *R. conorii* appears to be the ixodid ticks (primarily *Rhipicephalus* spp.) of rodents inhabiting the velds, forests, and semideserts of sub-Saharan Africa, but there is unequivocal evidence of *R. conorii* infection in wild rodents only in South Africa and Kenya.[44,59] Rodents elsewhere (India, Pakistan, and the Middle East) are undoubtedly infected, but little work has been done on them.**

*Considerable confusion about the identity and relationships of South African tick typhus existed in the 1930s.[34–38] Mason and Alexander finally resolved the problem by showing that both murine typhus (*R. typhi*) and a tick typhus rickettsia were present and responsible, respectively, for South African fleaborne (endemic) typhus and South African tick-bite fever. They suggested a system of nomenclature based on varietal names; thus the fleaborne typhus agent became *R. prowazeki* var. *mooseri* and the tick-bite fever agent, *Dermacentroxenus rickettsii* var. *conori*. They also stated that the generic name *Rickettsia* could not be allowed for *R. ruminantium, R. bovis, R. ovina*, or *R. canis*, although they conceded that they belong in the Family Rickettsiaceae.

**Wild rodents, bats, insectivores, carnivores, and primates in West Africa are reported to have agglutinating antibody to *R. conorii*,[66a] but the validity of these results has not been established. Strains of *R. conorii* have not been recovered from these animals in that region.

Table 2
KNOWN OR REPORTED ARTHROPOD HOSTS OF *RICKETTSIA CONORII*

Arthropod	Distribution of Arthropod[a]	Reported isolations
Amblyomma cohaerens Donitz[b]	East, Central Africa	42[c]
A. gemma Donitz	East, Central Africa	43[d]
A. hebraeum Koch	Southern Africa	44
A. variegatum (Fabricius)	Ethiopian Faunal Region West Indies, Cape Verde Is.	42, 45
Boophilus decoloratus (Koch)	Ethiopian Faunal Region	42,[c] 46, 47[e]
Dermacentor auratus group	Southeast Asia	48[f]
D. marginatus (Sulzer)	Europe, Russia	49–51
D. pictus Hermann[g]	Europe	52
D. reticulatus Koch	Europe, Northern Asia	50[h]
Haemaphysalis bispinosa Neumann	Asia	53[i]
H. leachii (Audouin)	Africa	44,[j] 45, 56
H. papuana Thorell	Southeast Asia	48
H. punctata Canestrini and Fanzago	Europe, N.W. Africa, Asia, Russia, Japan	51
H. sp.[k]	Southeast Asia	57
Hyalomma aegyptium (Linneaus)	Mediterranean, Near East Black Sea Region, South Africa	44, 58
H. albiparmatum Schulze and Schlottke	Kenya, Tanganyika, Togo	59
H. marginatum Koch	South Russia, Asia Mideast, North Africa	42[c]
Ixodes granulatus Supino	Southeast Asia	57[m]
I. hexagonus Leach	Europe	50,[h] 60[n]
I. ricinus (Linneaus)[o]	Europe, West Russia, India	51, 56, 60, 61[n]
Rhipicephalus appendiculatus Neumann[o]	Africa south of Sudan	44
R. evertsi Neumann	Ethiopian Faunal Region	42,[c] 44, 59
R. haemaphysaloides Supino	Oriental Faunal Region	48
R. pulchellus Neumann	Ethiopian Faunal Region	42[c]
R. sanguineus (Latreille)[p]	Cosmopolitan	57, 64, 65
R. simus Koch	Ethiopian Faunal Region	3, 6, 28,[q] 42, 45, 66[q]

[a]Tick distribution mainly from Hoogstraal[40] and Anastos.[41]

[b]The adults are primarily parasites of the African wild buffalo, *Syncerus caffer;* the nymphs are found on ground birds and tree rats, *Thammomys* spp. Cattle are often infested with this tick when on or near buffalo grazing grounds.

[c]Evidence for *R. conorii* infection in these ticks was the appearance of low-level complement-fixing antibodies in guinea pigs into which they were injected or on which they were allowed to feed. Most of the guinea pigs that recovered were partially or completely resistant to challenge with virulent *Rickettsia rickettsii.*

[d]A number of *Amblyomma gemma, A. variegatum* and *A. cohaerens* ticks that were positive for rickettsial agents in the hemolymph contained rickettsiae identified as belonging to the spotted fever group by fluorescent antibody staining.

[e]Gear and Bevan[47] state that tick typhus in South Africa can be transmitted by *Boophilus decoloratus, Rhipicephalus appendiculatus,* and *Amblyomma hebraeum.* These ticks were said to be veld-dwellers and were not found in houses or on domestic animals. The authors were probably using the term domestic animal to refer to dogs and cats rather than to domesticated herbivores since these ticks are common parasites of cattle.

[f]Presumptive isolation of a *Rickettsia* sp. in guinea pigs inoculated with tick suspensions is reported. The authors consider them to be strains of scrub typhus rickettsiae, but it would seem more likely in retrospect that they may have been strains of tick typhus, or perhaps (less likely) of Q-fever. See text for discussion.

Table 2 (continued)
KNOWN OR REPORTED ARTHROPOD HOSTS OF *RICKETTSIA CONORII*

[g]Called *Dermacentor reticulatus*, a synonym for *D. pictus* according to Anastos.[41]
[h]Antibodies to *R. conorii* were produced by pools of *D. reticulatus* and *I. hexagonus* inoculated into laboratory animals.
[i]Mesnard and Toumanoff[53] report presumptive isolation of a *Rickettsia* from this tick on the basis of clinical signs and symptoms produced in guinea pigs and the observation of rickettsia-like organisms in tissue smears. They consider the rickettsia to be the agent of "Tropical Typhus", that is, tsutsugamushi fever, but it seems more likely to have been a tick typhus rickettsia, if indeed it was a rickettsia at all.
[j]Gear and de Meillon[54,55] experimentally demonstrated transmission of *R. conorii* by *H. leachii* and reported transovarial transmission as well.
[k]Probably *H. semermis*, *H. papuana*, or both.
[l]According to Hoogstraal,[40] *H. aegyptium* does not occur in Tropical or Southern Africa. The actual identity of the *Hyalomma* in question is not known.
[m]The presumptive tick typhus isolates from Southeast Asia are assumed to be *R. conorii* for the purposes of this table, but it is by no means certain that they are actually strains of this rickettsia. See text for discussion.
[n]Numerous pools of *I. ricinus* and *I. hexagonus* ticks collected in France contained morphologically typical and atypical rickettsia-like organisms. They could not be readily adapted to chick embryos or laboratory animals, but produced antibodies to *R. conorii*, *C. burnetii*, and *Chlamydia psittaci*.
[o]Pijper and Dau[35] reported successful attempts to transmit *R. conorii* experimentally with *I. ricinus* and *R. appendiculatus*.
[p]Experimental transmission was demonstrated by Neitz et al.[62] and Philip.[63] Transovarial transmission was reported by Blanc and Caminopetros.[64]
[q]One *R. simus* was removed from a woman 7 days before she developed the first symptoms of tick typhus infection. A strain of *R. conorii* was isolated from the tick, strongly suggesting that it transmitted the infection to the woman.

Hoogstraal et al.[67] and Ormsbee et al.[68] reported spotted fever group agglutinins in 2 to 5% of more than 1000 serum samples from wild mammals in Egypt. No rickettsiae were isolated.

2. *R. conorii in India*

The earliest recorded case of human infection with *R. conorii* in South Asia was probably the attack of "spotted fever" suffered by Megaw after being bitten on the neck by a tick in the Himalayas in 1916.[69] The tick typhus rickettsiae that have been isolated in India appear to be antigenically identical to *R. conorii* of southern Europe and Africa,[12,18,74] although Philip et al.[22] and Kalra[71] originally thought they differed slightly from the agent of *fievre boutonneuse*. An isolate from ticks in West Pakistan is also closely related or identical to *R. conorii*.[19] Little is known of the basic ecology of tick typhus in the Indo-Pakistan Region. The only ticks known or suspected to be naturally infected in India are *Haemaphysalis leachii*, *Ixodes ricinus*, and *R. sanguineus* (possibly also *R. turanicus*). They are all parasitic on domestic animals, and none are indigenous to India. Thus, the introduction of *R. conorii* into India and Pakistan in domestic animal ticks from the west in historical times is a distinct possibility.

3. *Tick Typhus Strains in Southeast Asia*

The identity of the tick typhus strains in Southeast Asia and their relationship to *R. conorii* has not been established. The strains thus far isolated from wild vertebrates and ticks do not appear to be identical to African *R. conorii* or to Asian *R. sibirica;* the one from Thailand is distinct from either. They may represent a separate evolutionary line of rickettsiae which have developed in *Haemaphysalis* ticks, whose center of origin and distribution is in Southeast Asia.

4. Dispersion from Center of Origin

If, as seems likely, *R. conorii* evolved or differentiated in Africa, it may have extended its range north into Europe and east into India in recent times. The domestication of herbivores and canines not only provided large populations for ticks to feed on, but also made it possible for them and their symbiotes to travel long distances as man and his herds migrated from region to region. Thus, these now widespread domestic animal ticks of African origin may have played two roles in the evolution and distribution of *R. conorii:* (1) as indigenous African ticks in which the rickettsia arose and (2) as domestic animal ectoparasites in which it extended its range. Before the appearance of man, these ticks undoubtedly parasitized a wide range of large and small mammals, large ground birds, and to some extent reptiles, much as they still do today. Man's arrival and his eventual domestication of herbivores and canines brought them into other niches where they have become established. In recent times, *R. sanguineus,* as an extreme example, has been transported all over the world by man and his canine companion. The migrations were slow and over short distances at first, perhaps enabling both tick and rickettsia time to become established in native fauna outside their original home in the Ethiopian Faunal Region. In such a way, it may have spread into the southwestern Palearctic Region. Later migrations were faster and covered longer distances. In these more recently invaded regions, *R. sanguineus* has become established, but *R. conorii* has failed to do so. This suggests that *R. sanguineus* may not be a natural maintaining host of *R. conorii,* that suitable wild vertebrate hosts for the tick or the *rickettsia* are not present, or that the niche is already filled by other organisms. In any event, natural cycles of infection have not become established outside the Ethiopian Faunal Region and a few areas to the east. The distribution of *R. conorii,* although far smaller than that of *R. sanguineus* remarkably resembles that of the rhipicephalids as a whole (i.e., the Ethiopian Faunal Region). Most of the 70 species of known *Rhipicephalus* ticks are confined to continental Africa: *R. sanguineus* is cosmopolitan, two are Oriental, and five range through southern Europe and Russia.[72]

Although capable of hereditary transmission in *R. sanguineus,*[64] *R. conorii* apparently is not a true symbiote as is *Wolbachia persica* in *Argas arboreus,* for example, since it is not present in all individuals. A true symbiote should be transmitted to every one of its host's offspring, not just to a few of them as seemed to be true for *R. rickettsii* in *Dermacentor andersoni*[70] or *D. variabilis.*[71] However, Burgdorfer's application of the fluorescent antibody technique to this problem has produced data that discredit earlier views. He was able to demonstrate transovarial transmission of *R. rickettsii* to almost 100% of the progeny of *D. andersoni* females infected with a virulent laboratory strain of rickettsia. The F_1 progeny also passed the rickettsia to 100% of their eggs and the second filial generation larvae. The highest infection rate ever demonstrated in wild *D. andersoni* by classical isolation methods is 13.5%.[73] The fluorescein-tagged antibody technique, unfortunately, has not been applied to the natural tick hosts of *R. conorii.*

Distribution of *R. conorii* beyond the Ethiopean Region in recent times via domestic animal ticks is an attractive hypothesis and undoubtedly has occurred to some extent. The meager data available suggest that it reached India that way, but one cannot rule out the possibility that it was introduced into India in prehistoric times via ticks and rodents (such as *Tatera* and *Meriones*) of Middle East origin. In Volume I, Chapter 3 it was pointed out that western India contains a few faunal elements originating in Africa or the Middle East. It was also stressed that the Indian subcontinent has a relatively unique rodent fauna primarily of Southeast Asian origin, which has been isolated for a very long time. The same can be said for its ticks. The rich *Haemaphysalis* fauna of India has clear affinities with Southeast Asia, but only a few species are common to both regions. Because most Indian rodents and ticks are of Southeast Asian origin, one might also expect to find an indigenous tick typhus rickettsia resembling more closely the Southeast Asian form than the African. Unfortunately, only a few strains have been definitely isolated from Southeast Asia, and their antigenic relationship to *R. conorii* or *R. sibirica* has not yet been determined; however, the TT-118 strain isolated in Thailand[77] appears to be antigenically distinct from both *R. conorii* and *R. sibirica*.[19] The *R. conorii* isolate from West Pakistan was recovered from a pool of unidentified nymphal *Rhipicephalus* species ticks collected from the wild burrowing rat, *Nesokia indica*. If these ticks represent an indigenous species of *Rhipicephalus*, natural wild rodent-tick cycles of *R. conorii* must be considered possible. An alternative explanation is that, if such a cycle exists, it is secondarily derived from ticks that are primarily parasites of domestic animals and which introduced the rickettsia into the region in recent times.

B. *Rickettsia sibirica*

North Asian tick typhus (= Siberian tick typhus caused by *Rickettsia sibirica*—*Dermacentroxenus sibiricus* in the Russian literature) occurs throughout western, central, and eastern Siberia and is focally distributed in Central Asia, Mongolia, Kazak, Kirgisz, Armenia, the Maritime Territory, and Islands in the Sea of Japan. Early Russian work on Siberian tick typhus is reviewed by Crocker et al.,[78] and the ecology of *R. sibirica*, including discussion of its wild vertebrate hosts, by Zdrodovsky and Golinevich[65] and Hoogstraal.[79] Eighteen species of ixodid ticks have been reported to be naturally infected (Table 3), but the known distribution of the rickettsial agent roughly coincides with the distribution of only six species—*Dermacentor nuttalli, D. silvarum, D. marginatus, D. pictus, Haemaphysalis concinna,* and *H. japonica*. The range of the other species extends beyond that of *R. sibirica*. *Haemaphysalis japonica* extends into Korea and eastern China where *R. sibirica* probably exists, but there are no reports in the western literature of relevant studies in this vast region. *Dermacentor marginatus* ranges into Europe where *R. sibirica* may be replaced by other spotted fever group symbiotes or parasites.

The implication is that this organism is primarily a parasite of the indigenous *Dermacentor* and *Haemaphysalis* ticks of north and central Asia and has come to infect other, nonindigenous ticks secondarily. The adult stages of all the ticks from which *R. sibirica* has been isolated are parasites of domestic animals. This undoubtedly accounts for the great amount of work on these ticks in the Russian literature since they are also the prime vectors of North Asian tick typhus to man. However, they are also natural parasites of wild herbivores and, in some cases, of carnivores, and the larval and nymphal stages are found naturally on a wide range of small wild mammals and even birds. Serological studies in northern Asia demonstrated antibody to *R. sibirica* in several species of wild rodents and birds,[100,101] and a strain of *R. sibirica* has been isolated from tissues of wild birds,[101] suggesting the existence of basic natural cycles of infection in wild life.

1. Southern Limits

The range of *R. sibirica* extends south of Russia into West Pakistan where several strains were isolated from pools of unidentified *Rhipicephalus* and *Dermacentor* ticks as well as from *R. sanguineus, Hyalomma detritum,* and *H. marginatum*.[102] Infected ticks were collected in a variety of habitats varying from lowland plains to alpine meadows and mountain deserts. The range of *R. sibirica* overlaps that of *R. conorii* in the lowland plains region and that of a newly discovered rickettsia in the plains and mountain deserts of West Pakistan. The new rickettsia serotype is antigenically distinct from *R. sibirica* and *R. conorii* as well as from the Thai agent although there is strong one-

Table 3
KNOWN OR REPORTED ARTHROPOD HOSTS OF *RICKETTSIA SIBIRICA*

Arthropod	Distribution of arthropod[a]	Reported isolations	Transovarian transmission
Dermacentor marginatus (Sulzer)	Kazakhstan, Kirgizia, Iran, Altai Mountains to Central Europe[b]	25, 65, 81, 82[b]	65, 83
D. nuttalli Olenev[c]	Siberia, Central Asia, Mongolia, Tibet	65, 85	65, 83
D. pictus Hermann	Kazakhstan, Kirgizia, Iran, Altai Mountains to Central Europe	65	65, 83
D. silvarum Olenev[d]	East Siberia, Mongolia, most of U.S.S.R.	86, 87	65, 83
Haemaphysalis concina Koch[e]	Temperate Eurasia-Atlantic Ocean to Sea of Japan, South U.S.S.R.	87, 88, 89,[f] 90	
H. japonica Nuttall and Warburton	Maritime Prov., Sea of Japan, Korea, East China	87, 91	92[g]
H. punctata Canestrini and Fanzago	Europe, Northwest Africa, Southwest Asia, Russia, Japan	65	65, 83
Rhipicephalus sanguineus (Latreille)	Cosmopolitan	92,[h] 93	
R. turanicus Pomerantz and Matikashvili	U.S.S.R.	94	
R. schulzei Olenev	Southeast U.S.S.R., Northwest China	92	
Hyalomma detritum Schulze	Asia, Mideast, North Africa, South Europe	94	
H. dromedarii[i] Koch	South U.S.S.R., Afganistan, Iran, Iraq, North Africa	94, 95, 97[j]	97
H. excavatum Koch	Asia, Mideast, North Africa, South Europe	94	
H. marginatum Koch	South Russia, Asia, Mideast, North Africa	94	
Ixodes apronaphorus Schulze	Palearctic Region	92[h]	
I. plumbeus Leach	Palearctic Region	92[h]	
I. ricinus Latreille	Palearctic Region	25, 98[k]	
I. persulcatus Schulze[l]	U.S.S.R. in Taiga	93, 99	
Ceratophyllus tesquorum	Palearctic	85[m]	

[a]Mainly after Anastos.[41]

[b]A new serotype, *R. slovaca*, has been described from Slovakia. It is closely related to *R. sibirica* and the two are difficult to distinguish. Golinevich[80] compared strains of *R. conorii* from *R. sanguineus* collected in the Black Sea littoral and *R. sibirica* from *D. nuttalli* in Siberia and found them closely related but distinguishable by complement-fixation test. He also reported that Kenya tick typhus rickettsia from *H. leachii* and Indian tick typhus rickettsia from *R. sanguineus* were antigenically distinguishable from *R. conorii* and *R. sibirica*. This claim has not been substantiated.

[c]Experimental transmission demonstrated by Merinov.[84]

[d]All four species of *Dermacentor* listed are parasitic on domestic animals (and in some cases wild ungulates and carnivores) in the adult stage. The larvae and nymphs occur on numerous wild rodents and other small animals.

[e]The adult stages of the *Haemaphysalis* ticks listed all parasitize domestic animals to a greater or lesser extent. They also are found naturally on wild herbivores. The larvae and nymphs parasitize wild rodents and other small mammals and, in the case of *H. punctata*, many birds as well. *Ixodes persulcatus* has a similar host range.

[f]Probably the same as reported by Somov et al.[88] The authors also reported isolations from gamasid mites and from the vole, *Microtus fortis*.

Table 3 (continued)
KNOWN OR REPORTED ARTHROPOD HOSTS OF *RICKETTSIA SIBIRICA*

[g]*Rickettsia sibirica* was transmitted to larvae from infected females, but the authors were not able to demonstrate that these larvae or the nymphs derived from them were able to transmit the organism by bite. They also showed by guinea pig inoculation that adults derived from larvae infected transovarially maintained the rickettsiae for 20 to 25 months.

[h]A number of authors in the Russian literature claiming isolation of *R. sibirica* from these ticks are quoted. They also state that the ability to transmit rickettsiae transstadially and transovarially was proved for most of these species, but no data are given.

[i]Reported as *Hyalomma asiaticum* by Proreshnaya and Rapoport.[95] According to Delpy[96] this is a synonym of *H. dromedarii* Koch.

[j]Kulagin et al.[97] reported the isolation of rickettsiae from *H. asiaticum* (=*H. dromedarii*) collected in southeastern Turkhmenia. The results of animal experiments suggested to the authors that it differed significantly from *R. sibirica* on the one hand and *R. rickettsii* on the other and they placed it in a new species, *Dermacentroxenus deserti pavlovskyi*. Antigenic comparisons were not made and the status of *D. deserti* is unclear. I suspect, however, that it is at most a variant of *R. sibirica* and is not a valid species.

[k]Spotted fever group rickettsiae were seen in *Ixodes ricinus* by fluorescent antibody test. They were not identified to species.

[l]*Ixodes persulcatus* is the principal vector of tickborne encephalitis virus in the Novosibirsk Oblast[99] and in many other areas of the U.S.S.R.

[m]A strain of *R. sibirica* was isolated from a pool of fleas of this species collected from long-tailed Siberian susliks.

way cross-reaction in the toxin neutralization test with the latter. It was isolated from pools of *R. sanguineus, R. turanicus,* and other unidentified *Rhipicephalus* ticks.

2. *Western Limits*

The western limit of *R. sibirica* is not precisely known. Reports of complement-fixing antibody to *R. sibirica* in a high percentage of wild mammal sera in Iran[103] are difficult to evaluate since the specificity of the test was not demonstrated. As shown earlier in this chapter, *R. conorii* extends (albeit secondarily) into the Caspian and Black Sea Basins of southwestern U.S.S.R. overlapping with *R. sibirica.*[104] Spotted fever group rickettsia in eastern Europe, long assumed to be *R. sibirica* or *R. conorii,* may not be either. Two strains of rickettsia isolated from *Dermacentor marginatus* ticks in Czechoslovakia[23] have recently been shown to be closely related to *R. sibirica* and *R. conorii* but antigenically distinct from either.[24] *Dermacentor marginatus* is a known vector of *R. sibirica* in Asian Russia. In the adult stage, it parasitizes large domestic animals (nymphs and larvae parasitize rodents) and ranges through Europe to Great Britain.[41] The Czechoslovakian strains of tick typhus (named *R. slovaca*) were isolated from adult ticks collected in the area surrounding two villages, and there is evidence that a wild rodent focus exists in the vicinity. Spotted fever group antibodies are common among rodents, insectivores, hares, and wild herbivores in Czechoslovakia.[25,27,29,81,105] It is reasonable to assume that these animals are naturally infected with *R. slovaca* rather than *R. sibirica,* but definitive data are lacking. If true, this focus must have existed for a long time in relative isolation from *R. sibirica* foci to the East and *R. conorii* foci to the South. It may have arisen from the same ancestral stock that gave rise to *R. conorii* and *R. sibirica.*

It is conceivable, however, that *R. sibirica*-infected ticks were imported into Czechoslovakia on domestic animals from eastern Russia in recent times. Since *R. sibirica* is passed transovarially in this tick,[65,106] it may remain infected for many generations without necessarily establishing a focus of infection in wild rodents. It may have become established in an isolated focus in the mountains and gradually diverged, acquiring new antigens different from those of *R. sibirica*. The extent of its range is not known, but it may be widespread throughout eastern Europe where it forms an intermediate serotype between *R. sibirica* and *R. conorii*.

3. Southeastern Limits

There is a regrettable gap in our knowledge of North Asian tick typhus in China, which makes a discussion of *R. sibirica* evolution more speculative than one might wish. Literally nothing is known of the ecology or even of the occurrence of tick typhus in this area: if information exists, it is not readily available to western scientists. We know that the Southeast Asian *Rickettsia* sp. extends at least into Thailand, and the isolation of a rickettsial agent from *Haemaphysalis bispinosa* off a sambur deer, *Cervus (Rusa) unicolor* Cuv., in Vietnam has been reported.[53] The Thai agent is antigenically distinct from *R. sibirica,* but the identity of the Vietnam agent has not been established.

The Annamite Mountain chain along the eastern edge of the Mekong Delta and extending north by northwest to the Shan Plateau of Burma and connecting with the Yunnan Plateau may have impeded migration of faunal elements northward from Southeast Asia, but it certainly was not a completely effective barrier. North of the Red Basin of Szechwan, the Tsinling Shan now forms a major divide across China. This great mountain chain with rugged peaks to 4000 m probably served as a more effective barrier to north-south faunal migrations, at least since the mid-Tertiary. It would be of the greatest interest to know how far into China the *Rickettsia* spp. of South and Southeast Asia extend, and where the true southern limits of *R. sibirica* are.

C. *Rickettsia rickettsii*

The present distribution of *R. rickettsii* is the New World and the Nearctic and Neotropic Faunal Regions. Although probably not the oldest species phylogenetically, it is perhaps not an accident that this member of the tickborne rickettsiae was the first to be isolated. The early epidemiological history of Rocky Mountain spotted fever (RMsf) was to be repeated many times with other vector-borne diseases. When man blunders into previously undisturbed forests, as he often has and still does, he inevitably interferes with natural host-parasite associations that have existed for eons. The consequences are often unpleasant and occasionally tragic. Before the development of a vaccine and antibiotics, RMsf claimed a number of lives, and it still causes considerable morbidity and some mortality in East and Southeast U.S.

1. Acarine Hosts

Some 20 species of ticks have been reported to be naturally infected with *R. rickettsii* and an additional 6 species are capable of experimental transmission (Table 4). However, only two species (*Dermacentor andersoni* and *D. variabilis*) in North America and one (*Amblyomma cajennense*) in South America are closely associated with the epidemiology of spotted fever. These ticks readily bite man and are efficient transmitters of *R. rickettsii*.

a. Dermacentor andersoni

The wood tick, *D. andersoni,* ranges throughout the mountainous region of western U.S. and southwestern Canada. It is not host specific, the larvae and nymphs feeding on a variety of rodents and other small mammals. The adults feed on large domestic and wild mammals and on man. Intermediate-size mammals—labomorphs, porcupines, badgers—are hosts of all three life stages.[139,153,154]

Table 4
KNOWN OR REPORTED ARTHROPOD HOSTS OF *RICKETTSIA RICKETTSII*

Arthropod	Distribution	Isolation	Experimental transmission	Transovarian transmission
Amblyomma americanum (Linnaeus)	South U.S., Mexico, Central America	107, 108	107, 110	110
A. brasiliensis Aragao	South America		111	
A. cajennense (Fabricius)	Texas, Central America, South America	112–115	114	114[a]
A. cooperi Nuttall and Warburton	South America	114	114[b]	
A. ovale Koch	Central America, South America	118		
A. striatum Koch	South America	119	7, 113, 119	
Dermacentor andersoni Stiles	Western North America	120[c]	120	120
D. marginatus Sulzer	Europe, U.S.S.R.		109	
D. nitens Neumann[d]	South America		115	
D. occidentalis Marx	California, Oregon	126	127[e]	
D. parumapertus Neumann	Southwest United States, Mexico	128		
D. variabilis (Say)	Central U.S., Southeast Canada, Mexico	108,[f] 129–134	109	
Haemaphysalis leachii (Audouin)	Africa	135,[g] 136,[f] 137	40[h]	
H. leporispalustris (Packard)	North America, Central America, South America	11, 135,[g] 141–143,[g/i] 144[g]	141	141[j]
Ixodes brunneus Koch	North America, Central America, South America	143[g]		
I. dentatus Marx	East U.S.	142, 143,[g] 144[g]		
I. cookei Packard	East North America	108,[f] 134		
I. loricatus Neumann	South America	114	114	
I. scapularis Say	Central and East U.S.	108,[f] 134		
I. texanus Banks	U.S.	108,[f] 134		
Rhipicephalus sanguineus (Latreille)	Cosmopolitan	145,[k] 146, 147[l]	146	
Ornithodoros nicollei Mooser	Mexico, South America	147	148	148
O. parkeri Cooley	Northwest U.S.		149	149
O. rostratus Aragao	South America		150	
O. rudis Karsch	South America			152[n]
Otobius lagophilus Cooley and Kohls	West North America	118[l]		

[a]Presumed transovarial transmission only, since rickettsiae were demonstrated by guinea pig inoculation of unfed larvae collected on the ground. Monteiro et al.[116] earlier failed to demonstrate transovarial transmission.

[b]Travassos[114] cites other references to papers describing original work.

[c]Isolation of *R. rickettsii*, demonstration of experimental transmission, and apparent transovarial transmission were first demonstrated by Ricketts.[120–125] In this work the tick was misidentified as *Dermacentor occidentalis*. Wolbach[119] repeated and confirmed Ricketts' work and cites references to earlier work. Many investigators subsequently have recovered *R. rickettsii* from this tick.

[d]*Anocenter nitens* or *Otocentor nitens* of some authors.

[e]Parker[127] states that *Dermacentor occidentalis* is an efficient vector in the laboratory but gives no details.

Table 4 (continued)
KNOWN OR REPORTED ARTHROPOD HOSTS OF *RICKETTSIA RICKETTSII*

[f]Rickettsiae were detected by the hemolymph test[137] and identified as spotted fever group rickettsiae, possibly *R. rickettsii*, by fluorescent antibody staining.[132] Four *Amblyomma maculatum* ticks also contained spotted fever group rickettsiae, but they were not specifically identified. There is a distinct possibility that the latter were *R. parkeri* and some of the other rickettsiae may have been *R. montana* or some other spotted fever group serotype. Burgdorfer[139] suggests that all reports of *R. rickettsii* from *A. americanum* may be a related serotype. A high percentage of *R. sanguineus* removed from dogs also contained rickettsiae which stained with fluorescein-tagged antibody to *R. rickettsii*, but the staining pattern was not uniform and they were subsequently identified as a new serotype of spotted fever organisms.[140]

[g]These ticks were removed from birds and produced serological evidence of infection with spotted fever group rickettsiae when inoculated into guinea pigs.

[h]Hoogstraal[40] states that *H. leachii leachii* is an efficient experimental vector of *R. rickettsii*, but does not cite original work.

[i]A strain of *R. rickettsii* was isolated from a pool of larval ticks collected from birds in the eastern U.S.

[j]Eggs from infected female *H. leporispalustris* produced inapparent infection when inoculated into guinea pigs as shown by their immunity to subsequent challenge with virulent *R. rickettsii*.

[k]Adult *R. sanguineus* ticks from dogs produced fever in rabbits and fever and minor scrotal reactions (in some cases) when inoculated into guinea pigs. The authors saw rickettsia-like organisms in tissue smears from the infected guinea pigs, but did not prove that the agent transmitted was *R. rickettsii*.

[l]The infected ticks were all collected around spotted fever patients.

[m]In one experiment Monteiro[151] reported transmission of *R. rickettsii* by *O. rostratus* 13 days after feeding on an infected guinea pig but were unable to effect transmission 28 days after an infective feed. On the basis of these results alone, *O. rostratus* cannot be accepted as being capable of biological transmission of *R. rickettsii*.

[n]*Ornithodorus rudis* (=*O. venezulensis* Brumpt) did not transmit by bite, but the rickettsiae survived in the tick for 343 days and was passed transovarially in one experiment.

b. Dermacentor variabilis

The American dog tick, *D. variabilis*, ranges primarily from the Great Plains region to the east coast, but it is also found in California, Idaho, Washington, and probably in other western states.[155] Voles and mice are the principal hosts of the larval and nymphal stages. The adults feed mostly on dogs, but also on other carnivores and some herbivores. Occasionally squirrels and lagomorphs serve as hosts of the immature and adult ticks.[139,154,156]

c. Amblyomma cajennense

Amblyomma cajennense appears to be the principal vector and maintaining host of *R. rickettsii* in Brazil[157] and possibly throughout South America. Its range extends from Argentina to extreme southernmost parts of the U.S. Among the several species of South American ticks known to be infected with *R. rickettsii*, *A. cajennense* is the most important vector of the disease to man. It parasitizes mainly equines, dogs, sheep, goats, wild rabbits, and opossums.[114,158] In many respects, such as host preference and morphology, it is similar to *A. americanum*. The pathway of transmission to man could easily have originated with native rabbits and opossums, since naturally infected ticks other than *A. cajennense* feed also on domestic animals and wild rodents, opossums, or rabbits. Brazilian rabbits *(Sylvilagus minensis)*, cavy *(Cavia apecea)*, and capybara *(Hydrochoerus capybara)* are all susceptible to infection, and experimentally infected capybaras have been shown to infect ticks fed on them.[114]

d. Amblyomma americanum

The lone star tick, *Amblyomma americanum*, has been implicated in the epidemiology of RMsf in the South and Southeast U.S., but its importance as a vector of *R. rickettsii* is not clear. Its range overlaps that of *D. variabilis*, the principal vector in that region, and there are only three reports of natural infection with spotted fever group rickettsiae.[107,108,132] The range of *A. americanum* extends to Canada in the North and Argentina in the South. The recognition that there are several distinct serotypes of spotted fever group rickettsiae in addition to *R. rickettsii* in North American ticks, and the lack of specific identification of the *A. americanum* rickettsiae also demands a cautious interpretation of the data. There is certainly no compelling evidence that *A. americanum* is as closely associated with the maintenance of *R. rickettsii* as are *D. andersoni* and *D. variabilis*.

e. Other Amblyommids

Amblyomma striatum on dogs also has been found naturally infected with *R. rickettsii* in Brazil. It and *A. brasiliensis* can transmit the rickettsia experimentally.

f. Haemaphysalis leporispalustris

The rabbit tick, *Haemaphysalis leporispalustris*, has been found naturally infected with *R. rickettsii* and can transmit it.[11,135,141–144] The geographic range of this tick is that of its principal hosts, *Sylvillagus* spp., which are widely distributed throughout North and South America. Ground-frequenting birds, especially ruffed grouse in the North, are also important hosts of *H. leporispalustris*, and individuals infected with spotted fever group rickettsia have been collected from birds in the eastern U.S.[143,144] Birds also are known to develop rickettsemias of long duration after inoculation with *R. rickettsii*[159] and may serve as disseminating hosts. The strains of rickettsiae isolated from *H. leporispalustris* are clearly closely related serotypes of *R. rickettsii*, but as a group can be distinguished from it.[20]

2. *Wild Vertebrate Hosts*

One problem that remained essentially unresolved for half a century was the proof that wild vertebrates were naturally infected with *R. rickettsii*. Now, there is considerable serological evidence of *R. rickettsii* infection in various wild and domestic mammals and birds in several areas of North America.[11,128,135,161–166] As early as 1937[112,113] opossums in South America were reported to be naturally infected, but it was not until 1954 that the first strain of *R. rickettsii* was isolated from a wild vertebrate in North America—a meadow mouse, *Microtus pennsylvanicus*.[167] Over a decade later a strain was isolated from *Sylvilagus floridanus* and several strains from *Peromyscus leucopus*, *Sigmodon hispidus*, *M. pennsylvanicus*, *Didelphis marsupialis*, and *Pitymys pinetorum*, all trapped in Virginia.[168,169] Burgdorfer et al.,[165] in a carefully conducted serological survey of small animals in the Bitterroot Valley, Montana spotted fever focus, found complement-fixing antibodies in many ground squirrels, chipmunks, and snowshoe hares. The validity of these serological results was confirmed by the isolation of *R. rickettsii* from *Lepus americanus*, *Citellus lateralis*, and *Eutamias amoenus*. Later, Burgdorfer et al.[170] demonstrated rickettsemia sufficient to infect ticks in experimentally infected meadow mice, ground squirrels, chipmunks, and snowshoe hares. Two other species of Leporidae, *Sylvilagus audubonii* and *Lepus californicus*, have been shown to be susceptible to experimental spotted fever infection and to circulate the rickettsia in their blood for varying periods up to 6 days.[171] These animals are important hosts of *Dermacentor andersoni*, *D. parumapertus* (jack rabbits), and *Haemaphysalis leporispalustris* (cottontails and hares).

a. **The Role of Lagomorphs in *R. Rickettsii* Ecology and Evolution**

The natural acarine hosts of *R. rickettsii* in North America are several species of *Dermacentor* and possibly *Haemaphysalis leporispalustris*. *Dermacentor* ticks are probably the basic hosts in which *R. rickettsii* originated in North Asia and in which it diverged from the Asian form as the ticks likewise diverged from their Asian ancestors. The present distribution of *R. rickettsii,* however, coincides not with the distribution of New World *Dermacentor* ticks, but with that of *H. leporispalustris.* This tick, in all stages, is largely restricted to several species of wild leporids and certain ground birds.[172] It is found naturally infected in the U.S. and spontaneously transmits the rickettsiae to rabbits.[173] *Haemaphysalis leporispalustris* has not been found naturally infected in South America, but it has not been studied there as intensively as in North America.

In an important paper, Jellison[174] mapped the distribution of human infection with Rocky Mountain spotted fever in the western U.S., showing that it coincided with that of Nuttall's cottontail (*Sylvilagus nuttallii*). This animal is an important host of the immature stages of *Dermacentor andersoni,* as well as *H. leporispalustris.* He also showed that the other three species of *Dermacentor,* all more or less implicated as vectors of *R. rickettsii,* each coincides with the distribution of a rabbit: *D. variabilis* and the eastern cottontail, *S. floridanus; D. occidentalis* and the Pacific Coast brush rabbit, *S. bachmanni,* and *D. parumapertus* and the black-tailed jack rabbit, *Lepus californicus.* The fact that *Sylvilagus* is one of the few North American genera represented in South America may be significant. Experimentally, however, cottontails appear to be less susceptible than many rodents to *R. rickettsii* infection and Burgdorfer et al.[175] consider them to be of minor importance in the ecology of the organism. This does not necessarily mean that they were not involved in its evolution and early distribution.

The basic natural cycle of *R. rickettsii* in North America clearly involves *Dermacentor* spp. ticks and wild rodents; the role of *H. leporispalustris* and lagomorphs is not so well established. In South America, there are indications that similar cycles exist, and there cottontail rabbits (*Sylvilagus*) may play a significant role. They and the rabbit tick, *H. leporispalustris,* could have been instrumental in introducing *R. rickettsii* into South America. The presumed absence of other rickettsiae with tick vertebrate cycles would have allowed it to spread readily to other ticks and vertebrates.

The role of domestic animals in cycles of *R. rickettsii* infections in South America is not clear, but there is little evidence to suggest that the large herbivores are an important link in the chain. Dogs and sheep, which are hosts of several important vector tick species, are susceptible to experimentally induced RMsf. The disease is mild and may be inapparent, but rickettsiae are shed into the blood stream and are thus available to be injested by ectoparasites.[5] In human spotted fever infections, however, the vector ticks probably acquire the organism from wild vertebrates and pass it directly to man. Transovarian transmission has been shown in the most common vector species, *A. cajennense,* and this makes domestic animals even less significant as essential members of the cycle. Foci of infection in South American wild vertebrates clearly exist.

Whether rodents or rabbits are the most important vertebrates in *R. rickettsii* ecology actually has little relevance now since species of both groups may be intimately involved. Although Jellison's original hypothesis that cottontails are more important than other mammals as maintaining hosts of *R. rickettsii*[174,176] has been questioned,[175] it might be instructive to briefly look at the Leporidae as potential elements in the origin and evolution of *R. rickettsii.*

According to Nelson[177] the terms hare and rabbit were first used to distinguish the European species, *Lepus timidus* and *Lepus cuniculus* (now *Oryctolagus cuniculus*). The terms now have broader meaning: all members of the circumpolar genus *Lepus* belong to the hares, while the other genera of both the Old and New Worlds are rabbits. Hares live in forms and bear young with well-developed coats of hair and with their eyes open; rabbits live in burrows of one sort or another and produce naked young born with their eyes closed.

The Leporidae are distributed practically throughout the world, but are not native to Australia or most of the Oceanic Islands. Of the six Old World genera, only two, *Lepus* and *Oryctolagus,* are widely distributed. *Lepus* is found in parts of both Old and New Worlds. Distribution patterns of the others are limited and disjunct: *Pronolagus* in South Africa, *Nesolagus* in Sumatra, *Caprolagus* in the southern Himalayas, and *Pentalagus* in the Liu Kiu Islands off Japan.

The number and variety of the Leporidae appear to be greater in North America and lesser in South America than in any of the other continental areas. Two of the four North American genera, *Brachylagus* and *Romerolagus,* are peculiar to that continent; *Lepus* is circumpolar, and *Sylvilagus* occurs in both North and South America. *Sylvilagus* is divided into two subgenera: the typical *Sylvilagus* reaches its highest development in North America, and *Tapeti,* which has the greatest range of all American subgenera of rabbits extending from Virginia to Patagonia, reaches its greatest development in South America. Nelson[177] recognizes 15 New World species of *Lepus,* 14 of *Sylvilagus,* and 1 each of *Brachylagus* and *Romerolagus.*

The earliest reported lagomorph is represented by skeletal material from Mongolia in what are probably Paleocene desposits.[178] This material is not positively identified but is probably a lagomorph.[179] Fossil remains do not appear again until the upper Eocene of North America and Eurasia. Simpson[180] lists ancient Leporidae as important migrants from Eurasia to North America in the late Eocene. The Ochotonidae along with early cricetid rodents probably migrated from Eurasia to North America in the early Oligocene. Lagomorphs also migrated eastward in the late Pliocene and Pleistocene. The evidence thus suggests an Old World origin of the Lagomorpha, but it is far from conclusive.

Early systematists often included rabbits (hares and cottontails) in the Rodentia[181] and nonspecialists seldom separate the two orders. Although superficially resembling rodents, lagomorphs are actually quite different and are only remotely related phylogenetically.[182] The origin and affinities of the Lagomorpha are still moot questions; and, unless a more complete fossil record is obtained, they may never be answered to everyone's satisfaction. They appear to be an order at least as ancient as the Rodentia and are perhaps even older. Wilson[183] shows the Lagomorpha branching (hypothetically) from the late Cretaceous mammalian trunk prior to the insectivores. At this time Laurasia was separating from Gondwanaland, and conceivably, the lagomorph stem could have developed independently in the northern supercontinent, leaving the southern continental mass to the rodents. The Rodentia are thought to have evolved from the insectivore stem in the early Paleocene.

Whatever the origin of the Lagomorpha, they were clearly contemporaneous with the North American Eocene rodents, and the leporids may have accompanied them across the first Tertiary land bridge between Asia and North America. They diverged and radiated throughout the New World along with the rodents; new elements from Asia were added to the fauna in the early Oligocene, the late Pliocene, and Pleistocene. On the other hand, there is evidence that the present North American hares (*Lepus*), although originating in Asia, did not arrive in the New World until the Pleistocene.[180] The cottontails (*Sylvilagus*) probably evolved in North America and were widely distributed throughout both North and South America before appearance of *Lepus*. The subgenus *Tapeti* probably originated in North America as an offshoot of the same ancestor as the subgenus *Sylvilagus*. It ranged into Central and South America and became isolated for a sufficiently long time to develop its subgeneric characteristics. After the formation of the isthmus between North and South America, *Tapeti* migrated back into southeastern North America.

Rodents or cottontails or both were probably responsible for the introduction of *R.*

rickettsii (in ticks) into South America during the Oligocene or later. No South American rodents before the Oligocene are known.[179] The presence of *R. rickettsii* in South American amblyommids and the apparent absence in them of any agent of the *R. conorii* group may be taken as additional evidence that the Rickettsieae arose only after the placental mammals were well developed. This hypothesis would be on much firmer ground if the South American rickettsiae were more completely known. It does appear clear, however, that the spotted fever organism in South America is identical to that in North America and that it evolved in the North and was introduced into the southern continent in geologically recent times.

b. Origin and Dispersion

Rickettsia rickettsii is a branch of the same stem that produced *R. sibirica* in northern Asia. It was almost surely transported across the Bering Sea land bridges into North America at various times beginning with the first rodent migrations in the early Eocene. The affinities of New World placental mammals leave no doubt that these bridges were located between northern North America and northern and central Asia—the center of origin of *Dermacentor* and a major center of distribution of rodents. The isolation of New World mammal and tick fauna for long periods when the Bering Sea land bridge was submerged enabled them to diverge and develop into distinct taxa. So too, we can surmise, the rickettsiae in the early *Dermacentor* slowly diverged from the parent stem and eventually acquired the specific characteristics possessed by modern *R. rickettsii*.

The early migration routes led principally south along the western portion of North America and finally crossed the narrow channel separating North from South America. Minor routes also extended into the eastern portion of North America where the fauna became more or less isolated by the formation of the Rocky Mountain chain. The impression that RMsf has spread from the western to the eastern U.S. in the last few decades is probably not correct.[160] The rickettsia has been enzootic there for eons. The increasing human activity in natural foci in recent years has served to increase exposure to infection with the resultant increasing incidence of cases of spotted fever.

In South America the major tick vectors of *R. rickettsii* appear to be various species of *Amblyomma*. If the hypotheses as to the origin of *Amblyomma* in South America or Africa and the origin of *R. rickettsii* in northern Asia are correct, the present association of *R. rickettsii* with *Amblyomma* must be secondarily derived. South American *Amblyomma* ticks might have picked up the rickettsiae from animals infected by ticks derived from North America, since they are common parasites on many of the same animals. Unfortunately, we actually know very little about natural hosts or cycles of *R. rickettsii* in South America. Almost all study has been stimulated by and oriented towards elucidating the cycles in which man has become involved, and, consequently, most rickettsial isolations have been made from ticks parasitizing domestic animals. A number of naturally infected ticks are capable of transmitting the agent experimentally and some are reported to transmit it transovarially (Table 4). This might indicate that a long period of adaptation has taken place, but it does not necessarily mean that the association has evolutionary significance.

If there is an indigenous rickettsia in the *Amblyomma* ticks in South America, it might be more closely related to *R. conorii* of Africa than to *R. sibirica* or *R. rickettsii* of Asia and North America. This is not necessarily true, however, since *R. conorii* is mainly a symbiote of rhipicephalid ticks. It is more likely that an *Amblyomma Rickettsia* sp., should one exist in the ticks of South America and Africa, would be distinct from the other members of the family. *Rickettsia parkeri* in North American *Amblyomma americanum* ticks may be such an organism.

D. *Rickettsia parkeri*

The "maculatum agent", as *R. parkeri* was originally designated, was first isolated from *Amblyomma maculatum* ticks collected from cattle in Texas.[184] Additional strains were isolated from *A. maculatum* collected in Georgia[185] and Mississippi.[15] This south-central and southeastern region of the United States is also an enzootic area of *R. rickettsii* which is found primarily in *Dermacentor variabilis,* but also in *A. americanum, H. leporispalustris,* and several species of *Ixodes*. The maculatum agent has not been isolated from any of these ticks, nor has *R. rickettsii* been isolated from *A. maculatum*. So far as known, *R. parkeri* has not been isolated from any other ticks or from any other area of the world.

The reason for the apparent host specificity of these agents is not clear, but it might be related to an interference phenomenon. If a sufficiently large proportion of *A. maculatum* ticks are naturally infected with *R. parkeri,* any other rickettsia would be excluded from that particular niche. This could also explain the absence of *R. parkeri* in the natural acarine hosts of *R. rickettsii*. Price[186] gives a few observations of possible interference between low- and high-virulence strains of *R. rickettsii* in ticks, but no conclusions could be drawn from his limited study.

Amblyomma maculatum is a New World tick widely distributed in the southern portions of North America and throughout Central and South America. Thus, the absence of *R. parkeri* from Central and South America may be more apparent than real. Reports of attempted rickettsial isolation from this tick are few; isolations made from other ticks have all been reported to be *R. rickettsii*. However, the two species are not easily distinguishable and there could have been misidentifications. In the complement-fixation test, the one most frequently used for identifying rickettsial isolates, *R. parkeri*-immune guinea pig serum cross-reacts completely with *R. rickettsii* antigen.

Moreira and Magalhaes[112] and Travassos[113] made the interesting observation that rickettsial strains recovered from naturally infected wild opossums, *Didelphis marsupialis* and *D. aurita,* produced a milder disease in guinea pigs than virulent strains of *R. rickettsii* usually do. These may have been strains of *R. parkeri,* but Travassos[187] also reported that a virulent strain of Sao Paulo rickettsiae became less virulent after six serial passages in *D. aurita*. As a general rule, *R. parkeri* produces a much milder disease in guinea pigs than strains of *R. rickettsii* isolated from *Dermacentor* ticks in the U.S. However, low virulent strains of rickettsiae identified as *R. rickettsii* have been isolated from infected ticks, particularly *H. leporispalustris*.[127,188] The South American work is thus of doubtful significance.

On the basis of the toxin neutralization test[18] and the microimmunofluorescence test,[20] *R. parkeri* belongs to the major subgroup of tick typhus species which also includes *R. rickettsii, R. sibirica, R. conorii,* and two unclassified serotypes from *Dermacentor occidentalis* and *D. parumapertus*. It is most closely related to and forms a minor subgroup with *R. sibirica, R. conorii,* and the *D. parumapertus* agent.

E. *Rickettsia montana*

The Eastern Montana (or EM) agent, designated *Rickettsia montana* by Lackman et al.,[4] was initially isolated from *Dermacentor andersoni* and *D. variabilis* ticks collected in eastern Montana.[189] It also has been recovered from species of deer mice (*Peromyscus*) and meadow mice (*Microtus*) in the same focus[4] and from *Peromyscus leucopus* and *Pitymys pinetorum* in Virginia[169] and *D. variabilis* ticks in Ohio[137] and Massachusetts,[190] suggesting that it is widely distributed in natural cycles involving small rodents and ticks.

Rickettsia montana, although a natural parsite of the major tick hosts of *R. rickettsii* (*D. variabilis* and *D. andersoni*), is antigenically distinct from any of the tick typhus rickettsiae. In the microimmunofluorescence test, it is most closely related to *R. australis* and *R. rhipicephali* but is quite distinct from either.[20] Guinea pigs immunized with *R. montana* are almost completely protected against challenge with *R. rickettsii*.[191] Thus, it is tempting to consider *R. montana* as a low virulent variant of *R. rickettsii*, but the antigenic differences are such that specific status is justified. Biologically, *R. montana* also differs from *R. rickettsii* in that it is reported to replicate in the lumen of the louse gut as well as in the gut cells in which only a portion of the cytoplasm is invaded.[192] As with *R. canada* infection, the louse is not harmed by *R. montana* infection.

A similar agent, "Western Montana U", recovered earlier from *D. andersoni* ticks collected in the Bitterroot Valley, Montana, was considered to be a strain of *R. rickettsii* although it was completely avirulent for guinea pigs.[186] All strains of spotted fever rickettsiae are not equally pathogenic for the guinea pig (for example, the low virulent Iowa strain isolated by Cox,[193]), but seldom is one found to be completely and permanently lacking in virulence for this animal. The Iowa strain was initially inoculated into both embryonated eggs and guinea pigs. In guinea pigs it produced only mild signs of illness and was lost after several attempted guinea pig passages. However, the organism grew well in yolk sacs, and after a few passages became virulent for guinea pigs. After further yolk sac passages it reverted to avirulence—producing no illness in guinea pigs—but it conferred complete immunity to challenge with highly virulent *R. rickettsii*. Such strains, of course, cannot be isolated in guinea pigs; a more sensitive system, such as the developing chick embryo, must be used. For decades the guinea pig has been a standard laboratory animal for the isolation of rickettsiae; until relatively recently, few laboratories had made extensive use of embryonated eggs. Thus, we know little of the prevalence of such "avirulent" strains in ticks or their geographical distribution. Actually very little is known about the virulence of wild populations of any rickettsia species.

Virulence is a difficult term to define for the simple reason that it often refers not to a specific characteristic of the microbe but to a reaction caused by the microbe in a living host. Thus, the term is meaningless unless the host and the reaction it produces are specified. Virulence is usually measured by determining the least number of organisms that will produce a specified, recognizable reaction in a particular host. It may be death, febrile response, appearance of the organism in the blood, antibody production, or any of a range of reproducible clinical signs and symptoms. Standard laboratory host systems, by convention, are the most often used for determining virulence of microbial strains, and they provide the only means for comparison of strain virulence from one laboratory to another. However, in most cases the results of different laboratories are only approximately comparable; not only do techniques differ, but the pathogens being tested often have been subjected to widely different treatment; for example, the passage level may vary and the host system used for isolation or passage may differ from one laboratory to another. An isolate is often deliberately passaged through a susceptible host system until it produces a standard reaction, that is, until it attains a stable level of "virulence". It is sometimes practically impossible to evaluate the virulence of an agent isolated by one laboratory from natural hosts obtained in a particular locality and compare this virulence with that of the same species of microbe isolated by another laboratory in a different locality.

It is unreasonable to assume that a microbial population is completely homogeneous. Its individual components are doubtless as varied as those of any other population. In sampling any microbial population, we impose a powerful bias towards recovering that portion of the community of organisms that is capable of multiplying in the host system used in the isolation procedure. We may actually be imposing very potent selective pressures favoring certain subpopulations or organisms in the sample, and the strain that eventually "adapts" to the laboratory host system may not even be representative of the organisms in the original sample. This may not be so critical with organisms culturable in synthetic media, but it is of the greatest significance to obligate intracellular parasites.

Virulence can be validly used in reference to a particular host system and to a particular well-defined laboratory procedure only if the passage history of the microbial strain used is known and described. Ormsbee et al.[194] have attempted to do this by defining the limits of infectivity and pathogenicity of rickettsiae in defined host systems. Even when all these criteria are met, however, it should be recognized that the virulence of the particular strain may or may not be an accurate measure of the normal virulence of the population from which the strain was isolated. Even then, the data are still of only limited value in assessing natural host-parasite relationships since the highly inbred strains of animals used in the laboratory are not at all representative of genetically heterogeneous wild animal populations.

In the last analysis, the only way to determine the dynamics of host-parasite relations in a community is to study directly the effects of the microbe in question on members of the biota that are involved. A classic example of this type of approach is the work of Baltazard and his colleagues[195–197] on the ecology of plague in natural populations of *Meriones* and associated rodents in Kurdistan. Quan and Kartman[198] in California and Marchette[199] in Utah also used this technique in studies on the ecology of plague. In the Utah study, strains of *Yersinia pestis* isolated from fleas and rodents collected in a plague focus were found to be of low "virulence" for standard laboratory host systems. However, when laboratory-reared rodents of the same species that existed in the plague focus were tested, they proved as susceptible to the "low-virulence" strain as they were to so-called virulent plague. Or to put it another way, the virulence of the plague isolates varied from one wild species to another, but it did not differ significantly from that of standard, virulent laboratory strains in certain species.

These studies, although preferable to those that rely solely on standardized laboratory animals, still are not ideal. They are artificial in the sense that the bacilli were grown in an artificial medium and were administered by hypodermic needle and syringe. The influence of the natural vector, the flea for plague, was eliminated from the cycle. If there ever was any question about the importance of studying the entire host-parasite system rather than isolated components, it was laid to rest by the brilliant work of Price,[188] who showed, among other things, that a strain of *R. rickettsii* isolated from ticks collected from *Microtus pennsylvanicus* could not be passaged successfully in these mice. The strain died out after nine serial mouse-to-mouse transfers. However, when ticks were used, instead of a needle and syringe, to transfer the rickettsia from one mouse to the next, the strain could be maintained for at least 16 passages, after which the experiments were terminated.

The points made regarding virulence are relevant to the use (or misuse) of another common term—"reservoir". When a term legitimately employed by one discipline to describe a structure or process is borrowed by another discipline to describe something entirely different, difficulties in interpretation of exact meaning commonly arise. The term comes from the French *reserver*—to reserve—and is defined in the *Oxford English Dictionary* (3rd Ed.) as (1) "A receptacle of earthwork, masonry, etc. specially constructed to contain and store a large supply of water for ordinary uses. A place or area in which water naturally collects in large quantities. (2) A part of animal or plant in which some fluid or secretion is collected or retained. A part of some apparatus in which a liquid or fluid is contained. (3) Any receptacle for fluids (or vapours). A receptacle or repository for things or articles. A store or collection, a reserve supply of something.''

Thus, by classical definition, a reservoir is a static structure for holding inert materials (usually liquids) until needed or used. As such the term is completely inapplicable to a dynamic, living biological system. When used to describe a particular complex concept involving interactions of living organisms, it tends to convey the erroneous idea that one element of the system acts as a receptacle to store another element until it is used or until it is ready to interact with the other elements. The mycetomes in certain insects might be thought of as reservoirs containing supplies of symbiotic microorganisms that are used to infect the oocytes or eggs and ensure the continuity of the insect-symbiote relationship in succeeding generations. In this sense one could conceivably make a case for considering certain species of ticks as reservoirs of rickettsiae if they are the ultimate source of the microbe. This is exactly what I believe is true of the tickborne rickettsiae. However, no one would be so rash as to consider the tick as a mere receptacle for the storage of rickettsiae that it collects from other sources (e.g., vertebrates). The tick doesn't store them, *it grows them* and maintains them throughout its life span, during which it may pass them to a vertebrate and, through its eggs, to succeeding generations. Likewise, the mycetome is more than a mere storage bin for insect microbial symbiotes.

The term ''reservoir'', with its connotation of stasis is simply inadequate to describe the dynamic processes involved in rickettsia-tick-vertebrate cycles. At the very least, it should never be used as a noun, but rather as an adjective, such as, ''reservoir system'', to more accurately describe the situation that exists in nature. Better yet, the term should be discarded altogether and replaced by something else, such as ''primary host'' or ''maintaining host'' to indicate the faunal element in the ecosystem in which the rickettsia evolved and in which it may be maintained more or less permanently whether or not other faunal elements enter into the cycle in every generation. One could then designate other members of the cycle as secondary or tertiary hosts, depending on the intimacy of their relationship in the natural cycle. Baker[200] must have had something in mind when he developed his ''typical epidemic series''.

F. *Rickettsia rhipicephali*

The evidence that the brown dog tick, *Rhipicephalus sanguineus,* is a vector of *R. rickettsii* is largely circumstantial. In Mexico it is considered to be an important transmitter of spotted fever to man, but definitive proof is lacking. There is little question that it is an efficient experimental vector,[110] although it has never been found naturally infected in the U.S.[140] Anigstein and Bader[201] reported the isolation of a spotted fever rickettsia from *R. sanguineus* in Texas, but the organism produced only mild febrile and scrotal reactions in guinea pigs, and it was not rigorously identified as *R. rickettsii.*

It is now clear that *R. sanguineus* is naturally infected with an organism morphologically similar to spotted fever group rickettsiae,[202] but one that differs biologically and antigenically from *R. rickettsii.* Burgdorfer et al.,[140] using the hemolymph test,[138] detected rickettsia-like organisms in a large proportion of *R. sanguineus* removed from dogs in spotted fever foci in Mississippi and isolated the agent in tissue culture. Subsequent study showed it to be a distinct serotype in the spotted fever group. It is remotely related to the other recognized spotted fever serotypes by the microimmunofluorescence test[20] and has been proposed as a new species named *R. rhipicephali.*[203]

The brown dog tick is recognized as the major vector of *R. conorii* in the Old World, and it is tempting to speculate that *R. rhipicephali* is derived from it. Patients with boutonneuse fever characteristically exhibit a black eschar (*tache noire*) at the site of the tick bite. There is no evidence that boutonneuse fever occurs in the New World, but primary lesions resembling *tache noire* have been reported on patients in the U.S.[204] However, these may have been caused by infections with *R. akari*, which is not at all closely related to *R. conorii*.

A strain of *R. rhipicephali* has also been isolated from *D. variabilis* ticks in North Carolina.[203] Since this tick feeds on animals other than dogs, indigenous wild animal foci as well as tick-dog cycles may exist. If this is the case, *R. rhipicephali* is more likely to be a variant of *R. rickettsii* than of *R. conorii*. Its close relationship to *R. rickettsii* is shown by immunization-challenge experiments,[140,203] but similar studies have not been done with *R. conorii*.

G. *Rickettsia akari*

1. *Rickettsialpox*

Although almost surely not the original home of *R. akari*, rickettsialpox was first recognized as a distinct clinical entity in New York City in 1946 by two private physicians.[205,206] The account of this discovery and the subsequent investigation was brilliantly described by the renowned science writer B. Roueche.[207] The Public Health Service was alerted, and with remarkable speed, the clinical features were clearly described,[208] the etiologic agent was isolated from the blood of a patient, identified as a spotted fever-like agent, and named *R. akari*.[209]

A clue to the possible vector was the discovery by Charles Pomerantz, a pest control expert, of mites on the basement walls of one of the affected buildings containing rickettsialpox cases. Collections were made from several buildings and they were identified as *Allodermanyssus sanguineus* (=*Liponyssoides sanguineus*). A strain of rickettsia isolated from two pools of mites was shown to be identical to that isolated from rickettsialpox patients.[209] A strain of *R. akari* was later isolated from house mice trapped in the focus[210] and the epidemiology of the disease was thus complete.[208]

2. *Distribution*

The distribution of *R. akari* is peculiarly disjunct. It is found in eastern U.S.S.R., the Ukraine, Korea, and the U.S., predominantly on the eastern seaboard; however, cases of rickettsialpox have also been reported from Ohio and Utah.[211] The number of reported cases in the U.S. (mostly in New York City) has declined sharply from the early 1950s. By 1976, Shepard et al.,[212] at the Center for Disease Control, Atlanta, Georgia, had not seen or diagnosed a case for more than 20 years. In New York, which does its own laboratory diagnosis of rickettsialpox, cases continue to occur in low frequency.[213]

Epidemics of rickettsialpox in Ukrainian cities followed the same course as in New York City, and *R. akari* was isolated from patients and from *L. sanguineus* collected from mice and rats.[214] There is serological evidence for *R. akari* infection in humans in eastern Europe,[215,216] but no clinical cases of rickettsialpox have been definitively diagnosed. Rehacek[217] believes that *R. akari* probably also exists in Italy because the vector mite occurs there. Some clinical and serological evidence suggests the presence of rickettsialpox in South Africa.[44] Most of the patients whose infections resembled rickettsialpox apparently contracted them in the bushveld. In Gear's opinion, wild rodents, not house mice, are the South African maintaining hosts, and the vector may be some other mite than *Liponyssoides sanguineus*. Acceptance of this, however, must await isolation of the agent.

3. *Natural Hosts*

The etiological agent of rickettsialpox is clearly associated with the ubiquitous house mouse, *Mus musculus*, and the gamasid mite, *L. sanguineus*, in the U.S.[211] and in

Russia.[65] Hirst[218] originally described *L. sanguineus* from specimens collected from *Rattus rattus, Arvicanthus niloticus*, and *Acomys calurinus* in Egypt, but it was also identified in earlier collections made in Washington, D.C. in 1909. In the eastern U.S. it occurs mainly on *Mus musculus*. In Utah, the westernmost extent of its range in North America, specimens have been collected from *R. norvegicus*.[219]

The tropical rat mite, *Ornithonyssus bacoti*,* has been shown to be capable of transmitting *R. akari* experimentally, and transovarian transmission has been demonstrated,[220] but this mite has not been found naturally infected.

Rickettsia akari has not been isolated from wild rodents in Russia or the U.S., but its recovery from *Microtus fortis* in Korea[16] indicates that natural wild rodent cycles do exist. The only other rodent besides the house mouse found to be naturally infected is the domiciliated rat, *R. norvegicus*.[65]

4. *Affinities*

Rickettsia akari clearly belongs to the spotted fever group of rickettsia in the possession of the soluble group antigen, but it is distinguishable from all other rickettsia by a variety of serological and immunological tests. A relationship between *R. akari* and *R. australis* can be demonstrated in cross-protection tests in guinea pigs. However, they are clearly differentiated by species-specific CF antibody prepared in mice[4,17] and by the microimmunofluorescence test.[20] In the latter test, *R. akari* is not closely related to any known *Rickettsia*. The rickettsialpox rash is more similar to that of chickenpox than of RMsf,[208,223] and there is an eschar at the site of the bite of the infected mite. In the latter respect, rickettsialpox resembles boutonneuse fever, North Asian tick typhus, Queensland tick typhus, and scrub typhus.

At the molecular level, *R. akari* lacks one of the six major polypeptides shared by all other spotted fever group rickettsiae.[224] Ormsbee et al.[225] found cross-reacting antibody to *R. akari* in sera from both typhus and spotted fever cases, suggesting that it is related antigenically to typhus as well as to spotted fever rickettsiae.

5. *Summary*

The seemingly peculiar distribution of *R. akari* may not be peculiar at all. The picture simply may not be complete. It is a natural parasite of tiny gamassid mites, which most rickettsiologists ignore when looking for tickborne agents. If man had not interfered with the natural cycle by exposing his domiciliated mice to infection, and thereby secondarily exposing himself, we would not have heard of *R. akari* or even suspected its existence. When the necessary investigations are carried out, it will be found to range at least over much of the U.S.S.R., and perhaps even throughout northern North America. It will probably also be found, as now suspected, that the house mouse and *Liponyssoides sanguineus* have become secondarily involved by incidental association with field rodents and other gamassid mites that constitute the natural maintaining system for *R. akari*.

H. *Rickettsia australis*

Coincidentally with the discovery and description of rickettsialpox in New York City (and reminiscent of the Q fever story), Andrew et al.[226] described an outbreak of 12 cases of tick typhus-like illness in military personnel stationed in the Atherton tableland,

**Ornithonyssus bacoti* (Hirst) Bregetova, 1956, is also variously known in the literature as *Liponyssus bacoti* hirst, 1920; *Bdellonyssus bacoti* fonseca, 1941; *Macronyssus bacoti* Fonseca, 1941; and *Macronyssus bacoti* Buitendijk, 1945. It was originally described as *Leiognathus bacoti* Hirst, 1913, but this name has not been extensively used. In relevant medical literature, this mite is usually referred to as *Bdellonyssus bacoti*.[221,222]

North Queensland, Australia. Rickettsial agents isolated from two of the cases were shown to be distinct from *R. typhi*.[227] Study of one of these strains suggested that it resembled other tick typhus strains but could be distinguished from them immunologically.[8,14,17,228] Philip[229] named the organism *Rickettsia australis*.

1. Natural Hosts

Ths scrub tick, *Ixodes holocyclus*, was strongly suspected to be a vector of Queensland tick typhus on epidemiological grounds[226,230] and circumstantially because it is practically the only tick species that bites man in the endemic areas of Queensland.[231] Later *R. australis* was isolated from naturally infected *I. holocyclus* and *I. tasmani* in southeast Queensland,[232] and the existence of at least one and possibly two vector hosts was established. *Ixodes holocyclus* is widely distributed along the east coast of Australia from Cape York in North Queensland to Victoria. It is restricted to bush and scrub country but has a wide host range, feeding on all domestic animals, wild marsupials, and placental mammals including man.[233] *Ixodes tasmani* is also widely distributed throughout eastern and southern Australia where it feeds on a variety of small mammals.[232] Antibodies to *R. australis* have been demonstrated in a variety of bandicoots and rodents[234] which serve as hosts of these ticks.[235] Illness caused by *R. australis* is now recognized in southern Queensland and the earlier name, North Queensland tick typhus, has given way to simply Queensland tick typhus. An extensive survey might well necessitate a further name change to Australian tick typhus or eastern Australia tick typhus if *I. holocyclus* and *I. tasmani* are its only arthropod hosts.

Two haemaphysalids (*H. bancrofti* and *H. humerosa*) are also found in the Atherton Highland, but it is not known if they or other arthropods are involved in the natural cycle.

2. Affinities

Philip et al.[20] tentatively place *R. australis* in a minor subgroup of spotted fever group serotypes with *R. montana* and the *Ixodes pacificus* agent on the basis of the microimmunofluorescence test. They admit, however, that these serotypes are only remotely related to each other.

3. Origin

Rickettsia australis has not been reported outside Australia, and its northern limit is not known. The Indonesian and Philippine Archipelagoes may have been of great strategic importance to the evolution of *R. australis*, but the almost complete lack of knowledge concerning the tickborne rickettsiae of this huge region hinders our understanding of the evolutionary processes involved. Only one report[236] of a tickborne rickettsial agent in Indonesia is known to me. Guinea pigs inoculated with homogenates of *Dermacentor auratus*, *Rhipicephalus haemaphysaloides*, and *H. papuana* collected from wild boar in Sumatra developed fever. The rickettsiae seen in their tissues were thought to be *R. tsutsugamushi*, but may have been the Southeast Asian form of tick typhus. An intensive and extensive study of Indonesia and the Philippines would almost certainly reveal the presence of tickborne rickettsiae at least on the major islands. But where lies the boundary between the tick typhus *Rickettsia* sp. of mainland Asia and *R. australis* of Australia? And what is the nature of the boundary that has kept these forms isolated long enough for them to have diverged to the species we see today? These questions are still awaiting answers.

I. Newly Described Tick-Typhus Rickettsiae

1. *Ixodes pacificus* Strain

A spotted fever group rickettsia has been isolated on several occasions from the Pacific Coast tick, *Ixodes pacificus*, collected in Oregon.[237] All three strains tested were mildly virulent for guinea pigs and protected them against lethal *R. rickettsii* challenge. In this respect they resemble *R. montana*[189] and the low virulent western "U" strain of *R. rickettsii*.[188] They contain the spotted fever group-specific antigen as shown by cross-reactions with other members of the group in the CF test, but also possess specific antigenic determinants detected by microimmunofluorescence. The Oregon strains are most closely related antigenically to *R. montana* and *R. australis*, but even this relationship is remote.[20] The distribution of the *I. pacificus* strain is not known, nor is it known if other ticks harbor the agent or what mammals the natural cycle involves. *Ixodes pacificus* ranges along the Pacific Coast from British Columbia to southern California.[238] Similar to all *Ixodes*, it is a three-host tick and has a wide host range, feeding on rodents, rabbits, ground birds, lizards, domestic animals, and man.

2. *Dermacentor parumapertus* Strain

Strains of a spotted fever group rickettsia have been isolated from the rabbit dermacentor, *D. parumapertus*, in Nevada[239] and Utah.[128] Little is known about the ecology of this organism, but desert jack rabbits (*Lepus californicus*) are the major hosts of this tick. It ranges from the Columbia River in eastern Oregon south through California and southeast to western Texas. The organism is related but not identical to *R. rickettsii* in cross-immunity tests.[239] In the microimmunofluorescence test, it falls into a minor subgroup of spotted fever group organisms with *R. sibirica*, *R. conorii*, and *R. parkeri*.[20]

3. *Dermacentor occidentalis* Strain

Strains of a spotted fever rickettsia isolated from *Dermacentor occidentalis*, the Pacific Coast tick in California[240] are closely related to *R. rickettsii* but can be distinguished from it quantitatively by comparison of homologous and heterologous serum titers in the microimmunofluorescence test. These strains were isolated in Vero and mosquito cell cultures and produced no illness in laboratory animals although guinea pigs produced complement-fixing antibody to them. After four passages in tissue culture, they grew in eggs.

The Pacific Coast tick occurs in California and Oregon, where its range may overlap that of *D. variabilis* in the south and of *D. andersoni* in the north.[241] The larvae and nymphs feed on a variety of wild rodents and lagomorphs and occasionally on carnivores, domestic animals, and man.[242] Hosts of the adult stage include deer, rabbits, domestic animals, and man. The animals involved in the natural cycle of this rickettsia are not known.

4. *Hlp*-like *R. rickettsii* Strains

Parker et al.[11] isolated spotted fever group rickettsiae of low virulence from *H. leporispalustris* ticks and showed that they differed slightly from *R. rickettsii* in the guinea pig cross-immunity test. Later Bell et al.[189] recovered similar low virulent strains from *D. andersoni* and *D. variabilis* in Montana. Philip et al.[20] have separated *R. rickettsii* into high virulent, *R.*-like, and low virulent, *Hlp*-like strains. The *D. occidentalis* isolates and the *Hlp*-like isolates appear to be variants of *R. rickettsii*.[20] They differ slightly from virulent strains in guinea pig vaccination-challenge tests and in the microimmunofluorescence test.

5. *Amblyomma americanum* Strain

Other strains of spotted fever group rickettsiae which differ from *R. rickettsii* have been recovered from ticks in the U.S., but they are not yet characterized as to serotype. Strains from *Amblyomma americanum* collected in various parts of southeastern U.S.[134,140] react with fluorescein-conjugated *R. rickettsii* antibody, but are nonpathogenic for laboratory animals. Their pathogenicity for man is not known, nor is it clear if these strains are related to those isolated in 1943 from cases of Bullis fever in Texas. Bullis fever is a strange disease that occurred in epidemic form among the military at Camp Bullis, Texas, and was epidemiologically associated with tick (*A. americanum*) bites.[243] Sporadic cases may have occurred prior to 1943, but there have been no recognized outbreaks since the end of World War II.[244] Rickettsia-like agents were recovered from blood and lymph nodes from patients. The agents were isolated in guinea pigs, in which they produced a mild febrile illness, and maintained in mice and embryonated eggs.[245,246] They were unrelated by CF test to *C. burnetii* or *R. rickettsii*. Their etiological relationship to the disease in man was established serologically and by reproduction of the clinical syndrome in human volunteers. Identical strains were recovered from *A. americanum* ticks collected at Camp Bullis.[246-249] Antibody to the Bullis fever agent was demonstrated in deer and rabbits, natural hosts of *A. americanum*. It should be noted, however, that *A. americanum* carries *Francisella tularensis*,[250] *C. burnetii*,[156] and Lone Star virus.[251] Only *C. burnetii* was ruled out as an etiologic agent of Bullis fever.

The relationship of the Bullis fever agent to the *R. rickettsii*-like agents recently detected in *A. americanum* in the southeast[132,134] is not known. The latter agents were not infectious for voles and have not been isolated. The suggestion of Anigstein and Anigstein[244] that the Bullis fever agent be named *Rickettsia texiana* on the basis of retrospective analysis of the data is not justified. If strains of the organism still exist, they should be tested and compared to other tick isolates before they are named.

III. EVOLUTION OF THE *RICKETTSIA (DERMOCENTROXENUS)* TICKBORNE LINE

The *Dermocentroxenus* or tickborne line of rickettsiae represents the parent stem of the genus *Rickettsia* as defined here. It probably arose as an acarine symbiote that became peculiarly adapted to ixodid ticks as they diverged from the Mesostigmata and became obligate ectoparasites of vertebrates. As the spectacular reptile fauna diminished near the close of the Mesozoic, the emerging mammal fauna became the major hosts of ixodid ticks. Certain rickettsia-like symbiotes of some (but perhaps not all) ticks eventually developed a more complex life cycle involving homeothermous vertebrates as well as their traditional poikilothermic arthropod hosts. The present zoogeographical distribution of rickettsial species suggests that they arose in the Old World Tropics of Africa or Asia after the supercontinents of Gondwanaland and Laurasia split and separated from Pangaea about 150 million years ago, and the present land masses drifted apart. The apparent absence of truly indigenous *Rickettsia* spp. in South America is supporting evidence.

One hypothesis is that *Rickettsia* arose in the Asian Tropics as a primitive form of what is now *R. conorii* and *R. sibirica*. Westward migrations of the parent rickettsial form in ticks and mammals (probably rodents) and subsequent isolation in Africa resulted in the divergence and development of *R. conorii*, while northward migrations into central and North Asia, followed by partial isolation, produced *R. sibirica*. The present serotypes in Southeast Asia may represent the ancestral form, but more likely a third branch on the evolutionary tree.

An alternative hypothesis is that *Rickettsia*-like symbiotes were common to the most primitive ticks and evolved with them as they diverged on the developing mammals. A certain group of these symbiotes, ancestors of *Rickettsia*, possessed the biological potential for replicating in warm-blooded mammals. Perhaps only certain primitive ticks were infected with this ancestral *Rickettsia*—the *Rhipicephalis* organisms developed into the present-day *R. conorii* in Africa, the *Dermacentor* organisms developed into *R. sibirica* in Asia, and possibly the Southeast Asian form developed in *Haemaphysalis* ticks to form the serotypes that presumably are indigenous to that region.

Whatever the origin of *R. conorii* and *R. sibirica*, it seems clear that *R. rickettsii* diverged from *R. sibirica*. Long after *R. sibirca* had diverged from the main stem, but probably before it acquired all its present characteristics, tick- and rickettsia-bearing mammals from north and north-central Asia migrated across the Bering Sea land bridges into North America, became isolated, and culminated in the emergence of *R. rickettsii* as a distinct species. Later movements south across the Central American corridor brought it to South America where it remains today still indistinguishable from the North American form.

The ecological and biological processes leading to speciation among living organisms are a continuous process, and variants of *R. sibirica*, *R. rickettsii*, and probably *R. conorii* have arisen. Intensive study has identified some of these offshoots of *R. rickettsii* in the New World and of *R. sibirica* in the Old World. There is every reason to believe that future studies will uncover similar variants in Africa, Asia, and South America as well as additional ones in North America.

Rickettsia parkeri in North America might be a relatively recent variant of *Rickettsii*, but it seems more likely that it arose earlier and closer to the main *Rickettsia* stem as suggested by its closer relationship to *R. conorii* and *R. sibirica* than to *R. rickettsii* (at least in the microimmunofluorescence test). It is tempting to speculate that *R. parkeri* represents the *Amblyomma* line as *R. conorii* represents the *Rhipicephalus* line and *R. sibirica* and *R. rickettsii* represent the *Dermacentor* line of *Rickettsia*. If this is the case, there should be a variety of *R. parkeri*-like rickettsia in the rich *Amblyomma* fauna of Africa and South America. None has been reported, but intensive rickettsial investigations of the amblyommid fauna of these continents has not been made.

The origin of *Rickettsia akari* is difficult to derive. It is related to the other *Rickettsia* spp. morphologically and biologically and immunologically in possession of the group-specific antigen, but in many respects, as already described, its relationship is remote. There is little question that it is derived from a common ancestor that also gave rise to *R. conorii* and *R. sibirica*, but it must have diverged early, perhaps by adaptation to mesostigmatic mites. Or it remained associated with the Mesostigmata when the Metastigmata diverged from them. The ecology of *R. akari* is simply not very well known. The association with mites in the New World may be secondarily derived. There is no evidence that the eastern U.S. is the center of origin or distribution of *R. akari*. There is little doubt that it evolved in Asia, but whether in ticks or mites is not known.

The origin of *R. australis* may be easier to resolve. In Volume I, Chapter 4 on rodent evolution, it was shown how the placental mammals are thought to have come to Australia from Southeast Asia via Indonesian island stepping stones, the first murids probably arriving on driftwood rafts in the Miocene. Presumably ticks had preceded the placental mammals by many millions of years. The early reptilian fauna undoubtedly had its share of primitive acarine parasites (Australia currently has nine species of *Aponomma*[252]), as did the early marsupials. When the rodents came much later, they added a few more species to the sparse tick fauna. The primitive reptile ticks living in the Mesozoic Era doubtless had microbial symbiotes much as they do today, and each newly introduced tick also must have brought along its symbiotes. There are thus three possible pathways by which *R. australis* could have evolved:

1. From the symbiotes of indigenous reptile ticks that eventually adapted to marsupials
2. From symbiotes of ticks brought in by early marsupials
3. From the already evolved *Rickettsia* in the acarines and rodents that came in from Asia much later

The only other continent with an abundant living marsupial fauna is South America. It may be significant that it is also in the Southern Hemisphere and has been relatively isolated from the rest of the great land masses of the world for immense periods; apparently it has developed no indigenous tickborne rickettsiae pathogenic for vertebrates, at least none are known. The agent of South American tickborne spotted fever is essentially identical to *R. rickettsii* of North America and is easily derivable from it. It would appear that no indigenous acarine-rickettsia-vertebrate systems involving marsupials exist as they do for placental mammals. There also appear to be no such systems involving Rickettsiaceae and birds. We do not know why. These important biological problems are amenable to investigation, but the effort to resolve them has not been made.

For whatever biological reason, the primitive rickettsia-like symbiotes of the *Amblyomma* line of ixodid ticks presumably did not adapt to the marsupial and large placental hosts of these ticks. It is becoming more and more evident that the special relationship between acarine, rickettsia, and vertebrate is not a universal phenomenon in the animal kingdom, but one uniquely restricted to certain placental mammals and certain acarines. We can thus dispose of the hypothesis that the appearance of *R. australis* in Australia was in any way related to the late Mesozoic development and radiation of marsupials in South America or Australia. The most reasonable hypothesis is that *R. australis* reached Australia with the murids of Southeast Asia, but perhaps not in ixodid ticks. It may have come primarily in the gamassid mites parasitizing the early rodents that settled in Australia. *Ixodes holocyclus* and probably other Australian ixodids acquired the rickettsia secondarily and may or may not be the primary maintaining arthropods in the cycle. It may well be that gamassid mites are still the basic arthropod vectors and murid rodents the vertebrate hosts. A secondarily evolved cycle could have developed involving ixodid ticks and marsupials. The recovery of *R. australis* from certain species of small marsupials clearly indicates that they can become infected, but it has not been established that they are a necessary part of the basic maintenance system, or that they were involved in the introduction of *R. akari* to Australia.

This hypothesis assumes that the ancestral Old World *Rickettsia* gave rise to *R. akari,* which in turn diverged into *R. australis* in the southern continent. These two species represent geographic variants of the original parent that diverged very early from the primordial ixodid rickettsial symbiotes after they began parasitizing rodents. Gamassid mite ectoparasites of infected rodents presumably ingested rickettsiae which eventually became adapted to them, while the other line maintained its relationship with ixodids. When the rodents migrated north into Asia and south into Australia, they carried with them gamassid mites and the mite line of *Rickettsia* which eventually evolved into the two species *R. akari* and *R. australis,* respectively. Therefore, we might expect to find a gamassid mite rickettsia related to these two in Southeast Asian and Indonesian mesostigmatic mites.

REFERENCES

1. **Philip, C. B.,** Class III, Microtatobiotes, in *Bergey's Manual of Determinative Bacteriology,* 7th ed., Breed, R. S., Murray, E. G. D., and Smith, N. R., Eds., Williams & Wilkins, Baltimore, 1957, 931.
2. **Moulder, J. W.,** The Rickettsias, in *Bergey's Manual of Determinative Bacteriology,* 8th ed., Buchanan, R. E. and Gibbons, N. E., Eds., Williams & Wilkins, Baltimore, 1974, 882.
3. **Lackman, D. B. and Pickens, E. G.,** Antigenic types in the Rocky Mountain spotted fever group of rickettsiae, *Bacteriol. Proc.,* 3, 219, 1953.
4. **Lackman, D. B., Bell, E. J., Stoenner, H. G., and Pickens, E. G.,** The Rocky Mountain spotted fever group of rickettsias, *Health Lab. Sci.,* 2, 135, 1965.
5. **Badger, L. F.,** Rocky Mountain spotted fever: susceptibility of the dog and sheep to the virus, *Public Health Rep.,* 48, 791, 1933.
6. **Dyer, R. E.,** Relationship between Rocky Mountain spotted fever and "exanthematic typhus of Sao Paulo," *Public Health Rep.,* 48, 521, 1933.
7. **Parker, R. R.,** Observations on an infectious agent from *Amblyomma maculatum, Public Health Rep.,* 54, 1482, 1939.
8. **Plotz, H., Smadel, J. E., Bennett, B. L., Reagan, R. L., and Snyder, M. J.,** North Queensland tick typhus: studies of the aetiological agent and its relation to other rickettsial diseases, *Med. J. Aust.,* 2, 263, 1946.
9. **Davis, G. E. and Parker, R. R.,** Additional studies on the relationship of the virus of Rocky Mountain spotted fever and Sao Paulo exanthematic typhus, *Public Health Rep.,* 48, 1006, 1933.
10. **Davis, G. E. and Parker, R. R.,** Comparative experiments in spotted fever and boutonneuse fever, *Public Health Rep.,* 49, 423, 1934.
11. **Parker, R. R., Pickens, E. G., Lackman, D. B., Bell, E. J., and Thrailkill, F. B.,** Isolation and characterization of Rocky Mountain spotted fever rickettsiae from the rabbit tick, *Haemaphysalis leporuspalustris* Packard, *Public Health Rep.,* 66, 455, 1951.
12. **Bozeman, F. M., Humphries, J. W., Campbell, J. M., and O'Hara, P. L.,** Laboratory studies of the Rocky Mountain spotted fever group of rickettsiae, in Comm. Rickettsial Dis. Armed Forces Epidem. Bd., Symp. Spotted Fever Group Rickettsiae, Med. Sci. Publ. No. 7, Wissemann, C. L., Jr., Ed., Walter Reed Army Inst. Res., Washington, D.C., 1960, 7.
13. **Plotz, H., Bennett, B. L., Wertman, K., and Snyder, M. J.,** Cross-reacting antibodies in Rocky Mountain spotted fever, *Proc. Soc. Exp. Biol. Med.,* 57, 336, 1944.
14. **Lackman, D. and Parker, R. R.,** The serological characterization of North Queensland tick typhus, *Public Health Rep.,* 63, 1624, 1948.
15. **Lackman, D. B., Parker, R. R., and Gerloff, R. K.,** Serological characteristics of a pathogenic rickettsia occurring in *Amblyomma maculatum, Public Health Rep.,* 64, 1342, 1949.
16. **Jackson, E. B., Danauskas, J. X., Coale, M. C., and Smadel, J. E.,** Recovery of *Rickettsia akari* from the Korean vole, *Microtus fortis pelliceus, Am. J. Hyg.,* 66, 301, 1957.
17. **Pickens, E. G., Bell, E. J., Lackman, D. B., and Burgdorfer, W.,** Use of mouse serum in identification and serologic classification of *Rickettsia akari* and *Rickettsia australis, J. Immunol.,* 94, 883, 1965.
18. **Bell, E. J. and Stoenner, H. G.,** Immunologic relationships among the spotted fever group of rickettsias determined by toxin neutralization tests in mice with convalescent animal serums, *J. Immunol.,* 84, 171, 1960.
19. **Robertson, R. G. and Wisseman, C. L., Jr.,** Tickborne rickettsiae of the spotted fever group in West Pakistan. II. Serological classification of isolates from West Pakistan and Thailand: evidence for two new species, *Am. J. Epidemiol.,* 97, 55, 1973.
20. **Philip, R. N., Casper, E. A., Burgdorfer, W., Gerloff, R. K., Hughes, L. E., and Bell, E. J.,** Serologic typing of rickettsiae of the spotted fever group by microimmunofluorescence, *J. Immunol.,* 121, 1961, 1978.
21. **Philip, R. N., Paretsky, D., Weiss, E., and Wisseman, C. L., Jr.,** Rickettsiology conference, *J. Infect. Dis.,* 141, 112, 1980.
22. **Philip, C. B., Hughes, L. E., Rao, K. N. A., and Kalra, L.,** Studies on "Indian tick typhus" and its relation to other human, typhus-like rickettsioses, *Arq. 5th Int. Congr. Microbiol.,* Instituto Oswaldo Cruz, Rio de Janeiro, 1950, 571.
23. **Brezina, R., Rehacek, J., Ac, P., and Majerska, M.,** Two strains of Rocky Mountain spotted fever group recovered from *Dermacentor marginatus* ticks in Czechoslovakia. Results of preliminary serological identification, *Acta Virol. Engl. Ed.,* 13, 142, 1969.
24. **Urvolgyi, J. and Brezina, R.,** *Rickettsia slovaca:* a new member of spotted fever group rickettsiae, in *Proc. 2nd Int. Symp. Rickettsiae Rickettsial Dis.,* Kazar, J., Ormsbee, R. A., and Tarasevich, I. N., Eds., Slovak Academy of Sciences, Bratislava, Czechoslovakia, 1978, 299.

25. **Rehacek, J., Brezina, R., Zupancicova, A. M., and Kovacova, E.,** Contribution to the natural focality of rickettsiae belonging to the Rocky Mountain spotted fever (RMsf) group in Slovakia, *Folia Parasitol. (Prague),* 19, 41, 1972.
26. **Rehacek, J.,** Some aspects of ecology of rickettsiae, in, *Proc. 2nd Int. Symp. Rickettsiae Rickettsial Dis.,* Kazar, J., Ormsbee, R. A., and Tarasevich, I. N., Eds., Slovak Academy of Sciences, Bratislava, Czechoslovakia, 1978, 475.
27. **Rehacek, J., Urvolgyi, J., Brezina, R., Kazar, J., and Kovacova, E.,** Experimental infection of hare (*Lepus europaeus*) with *Coxiella burnetii* and *Rickettsia slovaca, Acta Virol. Engl. Ed.,* 22, 417, 1978.
28. **Yablonskaya, V. A.,** Identification of rickettsiae isolated from the brain of rodents trapped in the endemic areas in Czechoslovakia, in *Proc. 2nd Int. Symp. Rickettsiae Rickettsial Dis.,* Kazar, J., Ormsbee, R. A., and Tarasevich, I. N., Eds., Slovak Academy of Sciences, Bratislava, Czechoslovakia, 1978, 281.
29. **Rehacek, J., Urvolgyi, J., and Kovacova, E.,** Massive occurrence of the spotted fever group in fowl tampan, *Argas persicus,* collected in the Armenian S.S.R., *Acta Virol. Engl. Ed.,* 21, 431, 1977.
30. **Makarova, V. A., Tarasevich, I. V., and Plotnikova, L. F.,** Antigenic structure of rickettsiae isolated in Czechoslovakia and U.S.S.R. and their position in the spotted fever group, in *Proc. 2nd Int. Symp. Rickettsiae Rickettsial Dis.,* Kazar, J., Ormsbee, R. A., and Tarasevich, I. N., Eds., Slovak Academy of Sciences, Bratislava, Czechoslovakia, 1978, 293.
31. **Ormsbee, R. A. and Peacock, M. G.,** Antigenic relationships of rickettsiae of the typhus and spotted fever groups, in *Proc. 2nd Int. Symp. Rickettsiae Rickettsial Dis.,* Kazar, J., Ormsbee, R. A., and Tarasevich, I. N., Eds., Slovak Academy of Sciences, Bratislava, Czechoslovakia, 1978, 207.
32. **Burgdorfer, W., Aeschlemann, A., Peter, O., Hayes, S. F., and Philip, R. N.,** *Ixodes ricinus:* vector of a hitherto undescribed spotted fever agent in Switzerland, *Acta Trop.,* 36, 357, 1979.
33. **Freyche, M. J. and Deutschman, Z.,** Human rickettsioses in Africa, *Epidem. Vital Stat. Rep. W.H.O.,* 3, 160, 1950.
34. **Pijper, A. and Dau, H.,** Cross immunity between South African typhus and tick-bite fever, *Br. J. Exp. Pathol.,* 13, 33, 1932.
35. **Pijper, A. and Dau, H.,** Die Fleckfieberartigen Krankheiten des sudlichen Africa, *Zentralbl. Bakt. Parasitenkd. Infektionskr. Hyg. Abt. I Orig.,* 133, 7, 1934.
36. **Pijper, A. and Dau, H.,** South African typhus, *J. Hyg.,* 35, 116, 1935.
37. **Pijper, A. and Dau, H.,** Etude experimentale comparee de la fievre boutonneuse et de la tick-bite fever, *Arch. Inst. Pasteur, Tunis,* 25, 388, 1936.
38. **Pijper, A. and Crocker, C. G.,** Rickettsiosis of South Africa, *S. Afr. Med. J.,* 12, 613, 1938.
39. **Mason, J. H. and Alexander, R. A.,** Studies of the rickettsias of the typhus and Rocky Mountain spotted fever group in South Africa. IV. Discussion and classification, *Onderstepoort. J. Vet. Sci. Anim. Indust.,* 13, 67, 1939.
40. **Hoogstraal, H.,** African Ixodoidea. I. Ticks of the Sudan, Bureau of Medicine and Surgery, Department of the Navy, Washington, D.C., 1956, 1.
41. **Anastos, G.,** The Ticks or Ixodides of the U.S.S.R., Publ. No. 548, U.S. Public Health Service, Washington, D.C., 1957, 1.
42. **Philip, C. B., Hoogstraal, H., Reiss-Gutfreund, R. J., and Clifford, C. M.,** Evidence of rickettsial disease agents in ticks from Ethiopian cattle, *Bull. W.H.O.,* 35, 127, 1966.
43. **Burgdorfer, W., Ormsbee, R. A., Schmidt, M. L., and Hoogstraal, H.,** A search for the epidemic typhus agent in Ethiopian ticks, *Bull. W.H.O.,* 48, 563, 1973.
44. **Gear, J.,** The rickettsial diseases of southern Africa. A review of recent studies, *S. Afr. J. Clin. Sci.,* 5, 158, 1954.
45. **Heisch, R. B., McPhee, R., and Rickman, L. R.,** The epidemiology of tick typhus in Nairobi, *East. Afr. Med. J.,* 34, 459, 1957.
46. **Sampaio, A. and Faia, M. de M.,** A contribution to the study of the biology of *Rickettsia conorii* in Portugal, *Bol. Ins. Sup. Hig. Dr. Ricardo Jorge, Lisbon,* 7, 118, 1952.
47. **Gear, J. H. S. and Bevan, C.,** An outbreak of tick-bite (fever?), *S. Afr. Med. J.,* 10, 485, 1936.
48. **Kouwenaar, W. and Wolff, J. W.,** Onderzoehingen over sumatraansche Rickettsiosen, *Geneeskd. Tijdschr. Ned. Indie,* 74, 1659, 1934.
49. **Giroud, P., Capponi, M., and Dumas, N.,** Mise un evidence d'un agent du groupe des rickettsies sur *Dermacentor marginatus,* prevele sur bovin dans le centre de la France. Essai de modification du type, *Pathol. Microbiol.,* 24 (Suppl.), 104, 1961.
50. **Gilot, B.,** Recherches des rickettsies bebergees por les tiques (Acariens: Ixodoidea) du sud-est de la France. Premier bilan. Contexte ecologique de rencontre avec l'homme, *Bull. Soc. Pathol. Exot.,* 68, 529, 1975.

51. **Rehacek, J., Palanova, A., Zupancicova, M., Urvolgyi, J., Kovacova, E., Jarabeck, L., and Brezina, R.**, Study of rickettsioses in Slovakia. I. *Coxiella burneti* and rickettsiae of the spotted fever (SF) group in ticks and serological surveys in animals and humans in certain selected localities in the Lucenec and V. Krtis Districts, *J. Hyg. Epidemiol. Microbiol. Immunol.*, 19, 105, 1975.
52. **Giroud, P., Capponi, M., Dumas, N., and Colas-Belcour, J.**, De la fievre boutonneuse mediterraneene au groupe boutonneux pourpre, *Bull. Soc. Pathol. Exot.*, 56, 629, 1963.
53. **Mesnard, J. and Toumanoff, C.**, Agent infectieux du genre *Rickettsia* mis en evidence chez une tique (*Haemaphysalis bispinosa* Neumann) du cerf de la Cochinchine, *C.R. Acad. Sci.*, 210, 378, 1940.
54. **Gear, J. and de Meillon, B.**, The common dog tick, *Haemaphysalis leachi* as a vector of tick typhus, *S. Afr. Med. J.*, 13, 815, 1939.
55. **Gear, J. H. S. and de Meillon, B.**, The hereditary transmission of the rickettsiae of tick-bite fever through the common dog tick, *Haemaphysalis leachi, S. Afr. Med. J.*, 15, 389, 1941.
56. **Kalra, S. L.**, Progress in the knowledge of rickettsial diseases in India, *Indian J. Med. Res.*, 47, 477, 1959.
57. **Marchette, N. J.**, Rickettsioses (tick typhus, Q-fever, urban typhus) in Malaya, *J. Med. Entomol.*, 2, 339, 1965.
58. **Alexander, R. A., Mason, J. H., and Neitz, W. O.**, Studies of the typhus-Rocky Mountain spotted fever group in South Africa. I. Isolation of strains, *Onderstepoort J. Vet. Sci. Anim. Ind.*, 13, 19, 1939.
59. **Heisch, R. B., Grainger, W. E., Harvey, A. E. C., and Lister, G.**, Feral aspects of rickettsial infections in Kenya, *Trans. R. Soc. Trop. Med. Hyg.*, 56, 272, 1962.
60. **Giroud, P., Capponi, M., Dumas, N., Colas-Belcour, J., and Masson, R.**, Mise en evidence d'une facon presque constante sur les tiques de l' Est de la France de l'antigene du groupe boutonneux pourpre et isolement de souches, *C.R. Acad. Sci.*, 255, 611, 1962.
61. **Giroud, P., Capponi, M., Dumas, N., and Rageau, J.**, Les *Ixodes ricinus* et *hexagonus* de France contiennent des agents rickettsiens au proches, *C.R. Acad. Sci.*, 260, 4874, 1965.
62. **Neitz, W. O., Alexander, R. A., and Mason, J. H.**, The transmission of tick bite fever by the dog tick, *Rhipicephalus sanguineus* Latr., *Onderstepoort J. Vet. Sci. Anim. Ind.*, 16, 9, 1941.
63. **Philip, C. B.**, Tick transmission of Indian tick typhus and some related rickettsioses, *Exp. Parasitol.*, 1, 129, 1952.
64. **Blanc, G. and Caminopetros, J.**, Etudes epidemiologiques et experimentales sur la fievre boutonneuse, *Arch. Inst. Pasteur Tunis*, 20, 343, 1932.
65. **Zdrodovsky, P. F. and Golinevich, H. M.**, *The Rickettsial Diseases*, Pergamon Press, New York, 1960, 1.
66. **Timms, G. L., Heisch, R. B., and Harvey, A. E. C.**, Further evidence that *Rhipicephalus simus* Koch is a vector of tick typhus in Kenya, *East Afr. Med. J.*, 36, 114, 1959.
66a. **Le Noc, A., Richenbach, A., Ravine P., and Le Noc, D.**, Enquete serologique sur les rickettsioses animales au cameroun. II. Resultats de l'enquete, *Bull. Soc. Pathol. Exot.*, 70, 410, 1977.
67. **Hoogstraal, H., Kaiser, M. N., Ormsbee, R. A., Osborn, D. J., Helmy, I., and Gaber, S.**, *Hyalomma (Hyalommina) rhipicephaloides* Neumann (Ixodoidea: Ixodidae): its identity, hosts, and ecology, and *Rickettsia conori, R. prowazeki*, and *Coxiella burneti* infections in rodent hosts in Egypt, *J. Med. Entomol.*, 4, 391, 1967.
68. **Ormsbee, R. A., Hoogstraal, H., Yousser, L. B., Hildebrandt, P., and Attala, W.**, Evidence for extra-human epidemic typhus in the wild animals of Egypt, *J. Hyg. Epidem. Microbiol. Immunol.*, 12, 226, 1968.
69. **Megaw, J. W. D.**, A case of fever resembling Brill's disease, *Indian Med. Gaz.*, 52, 15, 1917.
70. **Zdrodovskii, P. F. and Golinevich, E. M.**, Differentiation of the causative organism of East African tick-borne rickettsiosis (Kenya tick typhus) and Indian tick-borne rickettsiosis (Indian tick typhus), *Prob. Virol. U.S.S.R.*, 3, 219, 1958.
71. **Kalra, S. L.**, Natural history of typhus fevers in India, *Indian J. Med. Sci.*, 6, 569, 1952.
72. **Walker, J. B.**, Some observations on the classification and biology of ticks belonging to the genus *Rhipicephalus*, with special reference to the immature stages, *East Afr. Med. J.*, 38, 232, 1961.
73. **Ricketts, H. T.**, Some aspects of Rocky Mountain spotted fever as shown by recent investigations, *Med. Rec. 1866–1922*, 76, 843, 1909.
74. **Price, W. H.**, The epidemiology of Rocky Mountain spotted fever. II. Studies on the biological survival mechanisms of *Rickettsia rickettsii*, *Am. J. Hyg.*, 60, 292, 1954.
75. **Burgdorfer, W.**, Investigation of "transovarial transmission" of *Rickettsia rickettsii* in the wood tick, *Dermacentor andersoni*, *Exp. Parasitol.*, 14, 152, 1963.
76. **Philip, C. B.**, Some epidemiological considerations in Rocky Mountain spotted fever, *Public Health Rep.*, 74, 595, 1959.

77. **Shirai, A., Bozeman, F. M., Humphries, J. W., Elisberg, B. L., and Faber, J. E.,** Experimental infection of the cotton rat, *Sigmodon hispidus* with *Rickettsia rickettsii*, *J. Bacteriol.*, 94, 1334, 1967.
78. **Crocker, T. T., Bennett, B. L., Jackson, E. B., Snyder, M. J., Smadel, J. E., Gauld, R. L., and Gordon, M. K.,** Siberian tick typhus. Relations of the Russian strains to *Rickettsia prowazeki*, *Public Health Rep.*, 65, 383, 1950.
79. **Hoogstraal, H.,** Ticks in relation to human diseases caused by *Rickettsia* species, *Annu. Rev. Entomol.*, 12, 377, 1967.
80. **Golinevich, H.,** A propos de la differenciation de quelques rickettsies du groupe de la fievre pourpree a tiques, *Arch. Inst. Pasteur Tunis*, 37, 13, 1960.
81. **Rehacek, J., Zupancicova, M., Ac, P., Brezina, R., Urvolgyi, J., Kovacova, E., Tarasevic, I. V., Jabolonskaja, V. A., Pospisil, R., and Baloghova, D.,** Rickettsiosis studies. II. Natural foci of rickettsioses in east Slovakia, *Bull. W.H.O.*, 53, 31, 1976.
82. **Brezina, R., Rehacek, J., Ac, P., and Majerska, M.,** Two strains of Rocky Mountain spotted fever group recovered from *Dermacentor marginatus* ticks in Czechoslovakia. Results of preliminary serological identification, *Acta Virol. Engl. Ed.*, 13, 142, 1969.
83. **Piontkovskaya, S. P. and Korschunova, O. S.,** Asian tick typhus, in *Human Diseases with Natural Foci*, Pavlovsky, Y. N., Ed., Foreign Languages Publishing House, Moscow, 1963, 98.
84. **Merinov, V. A.,** Investigation of North Asiatic (Siberian) rickettsiosis of *Dermacentor nuttalli* ticks collected in one of the Krasnoyarsk Region foci, *Med. Parasit. Moscow*, 1, 54, 1963.
85. **Korshunova, O. S., Petrova-Piontkowskaya, S. P., and Flint, V. E.,** Natural tick typhus fever foci in Baryat ASSR., *Zool. Zh.*, 44, 980, 1965.
86. **Bocharova, T. V.,** On the epidemiology of the tick spotted typhus, *Zh. Mikrobiol., Epidemiol. Immunobiol.*, 1–2, 68, 1943.
87. **Kulagin, S. M., Somov, G. P., Silich, M. I., Suvorova, L. V., and Bobrovskii, V. N.,** Further observations on tick-borne rickettsiosis in the Maritime Territory, *J. Microbiol. Epidemiol. Immunobiol. U.S.S.R.*, 31, 1652, 1961.
88. **Somov, G. P., Shapiro, M. I., and Petrov, A. A.,** Study of an island focus of northern Asian tick-borne typhus, *J. Microbiol. Epidem. Immunobiol. U.S.S.R.*, 29, 761, 1958.
89. **Shapiro, M. I.,** Experimental study of strains of tick-borne rickettsias isolated on one of the islands off the south coast, *J. Microbiol. Epidem. Immunobiol. U.S.S.R.*, 29, 1664, 1958.
90. **Yastrebov, V. K.,** Finding of natural infection of *Dermacentroxenus sibiricus* rickettsia in *Haemaphysalis concinna* Koch ticks in Altai Region, *Med. Parazit. Moscow*, 38, 105, 1969.
91. **Somov, G. P. and Shestakov, V. I.,** On spontaneous infection of *Haemaphysalis japonica douglasi* Nutt. and War. ticks with *Dermacentroxenus sibiricus* rickettsiae in Primorsk Region, *J. Microbiol. Moscow*, 40, 51, 1963.
92. **Belikova, N. P. and Somov, G. P.,** Study of ability of *Haemaphysalis japonica douglasi* Nutt. and War. and *Haemaphysalis neumanni* D. to assimilate rickettsiae under experimental conditions, *Dokl. Akad. Nauk SSSR*, 173, 981, 1967.
93. **Kotsinyan, M. E.,** Endemic rickettsioses in the Armenian S.S.R., *10. Sovesch-Parasitol. Prob. Moscow*, 1, 87, 1959.
94. **Zhmaeva, Z. M., Pchelkina, A. A., and Berdyev, A. B.,** Ixodid ticks as reservoirs and causes of mixed infections, *Med. Parasit. Moscow*, 38, 405, 1969.
95. **Proreshnaya, T. L. and Rapoport, L. P.,** Study of natural foci of tick rickettsioses in southwestern Kirgizia, *Zh. Mikrobiol. Epidemiol. Immunobiol.*, 40, 56, 1963.
96. **Delpy, L. P.,** Essai critique de synonymie du genre *Hyalloma* C. L. Koch 1844 depuis Linne, 1758, *Ann. Parasitol. Hum. Comp.*, 24, 464, 1949.
97. **Kulagin, S. M., Zhmaeva, Z. M., Shekhanov, M. V., and Pchelkina, A. A.,** Contribution to the characteristics of tick-borne rickettsiosis in southeastern Turkmenia, *Zh. Mikrobiol. Moscow*, 28, 114, 1957.
98. **Aeschlimann, A. and Burgdorfer, W.,** Aspects nouveaux du role et vecteur joue par *Ixodes ricinus* en Suisse, *Acta Trop.*, 36, 181, 1979.
99. **Netsky, G. I. and Shaiman, M. S.,** Distribution and relationships between foci of tick-borne encephalitis, tick typhus of northern Asia and Q-fever in Western Siberia, *Med. Parasit. Moscow*, 33, 136, 1964.
100. **Shayman, M. S., Stolbov, N. M., and Christyakov, A. A.,** Finding of complement-fixing antibodies to the agents of tickborne typhus fever in northern Asia and Q-fever in wild animals in extreme northwestern Siberia, *Med. Parazit. Moscow*, 37, 617, 1968.
101. **Somov, G. P. and Soldatov, G. M.,** The role of birds in the circulation of the tick typhus causative agent in nature, *Zh. Mikrobiol. Epidemiol. Immunobiol.*, 41, 126, 1964.
102. **Robertson, R. G., Wisseman, C. L., Jr., and Traub, R.,** Tick-borne rickettsiae of the spotted fever group in West Pakistan. I. Isolation of strains from ticks in different habitats, *Am. J. Epidemiol.*, 92, 382, 1970.

103. **Hamidi, A. N., Saadatezadeh, H., Tarasevich, I. V., Arata, A. A., and Fathangazad, A.,** A serological study of rickettsial infection in Iranian small mammals, *Bull. Soc. Pathol. Exot.*, 67, 607, 1974.
104. **Makarova, V. A.,** Antigenic structure of rickettsia of the tick-borne spotted fever group isolated in the Armenian S.S.R., *Zh. Mikrobiol. Epidemiol. Immunobiol.*, 55, 112, 1978.
105. **Rehacek, J., Palanova, A., Zupancicova, M., Urvolgyi, J., Kovacova, E., Jarabek, L., and Brezina, R.,** Study of rickettsioses in slovakia. I. *Coxiella burneti* and rickettsiae of the spotted fever (SF) group in ticks and serological surveys in animals and humans in certain selected localities in the Lucenec and V. Krtis Districts, *J. Hyg. Epidemiol. Microbiol. Immunol.*, 19, 105, 1975.
106. **Piontkovskaya, S. P. and Korschunova, O. S.,** Experiments of infection in *Dermacentor marginatus* Sulz. and *D. pictus* Herm. ticks with *Rickettsia prowazeki, Vopr. Kraj. Obshchestvo Exp. Parasitol.*, 8, 29, 1953 (in Russian).
107. **Parker, R. R., Kohls, G. M., and Steinhaus, E. A.,** Rocky Mountain spotted fever; spontaneous infection in the tick *Amblyomma americanun, Public Health Rep.*, 58, 721, 1943.
108. **Loving, S. M., Smith, A. B., DiSalvo, A. F., and Burgdorfer, W.,** Distribution and prevalence of spotted fever group rickettsiae in ticks from South Carolina with an epidemiological survey of persons bitten by infected ticks, *Am. J. Trop. Med. Hyg.*, 27, 1255, 1978.
109. **Maver, M. B.,** Transmission of spotted fever by the tick in nature, *J. Infect. Dis.*, 8, 327, 1911.
110. **Parker, R. R., Philip, C. B., and Jellison, W. L.,** Rocky Mountain spotted fever: potentialities of tick transmission in relation to geographical occurrence in the United States, *Am. J. Trop. Med.*, 13, 341, 1933.
111. **Travassos, J.,** Transmission experimentale du "typhus" exanthematique de Sao Paulo por l'*Amblyomma brasiliense* Aragao 1908, *C. R. Seances Soc. Biol. Paris*, 127, 1375, 1938.
112. **Moreira, J. A. and de Magalhaes, O.,** Typho exanthematico em Minas Geraes, *Bras. Med.*, 49, 465, 1935.
113. **Travassos, J.,** Resultats des inoculations experimental series due virus du "typhus exanthematique de Sao Paulo" a la sarique marsupiale (*Didelphis aurita*), *C. R. Seances Soc. Biol. Paris*, 126, 1049, 1937.
114. **Travassos, J.,** Studies on rickettsial diseases in Brazil, in *Proc. 4th Int. Congr. Trop. Med. Malaria*, Vol. 1, Department of State, Washington, D.C., 1948, 414.
115. **Patino-Comargo, L.,** Nuevas observaciones sobre un tercer foco di fiebre petequial (maculosa) en el hemisferio Americano, *Pan. Am. Union Bol. Off. Sanit.*, 20, 1112, 1941.
116. **Lemos Monteiro da Silva, J. L., Fonseca, F., and Prado, A.,** Typho endemico de S. Paulo. VI. Pesquesas sobre a possibilidade da transmissao experimental do virus por "Ixodidae," *Bras. Med.*, 46, 49, 1932.
117. **Salles Gomes, L.,** Typho-exanthematico de Sao Paulo, *Bras. Med.*, 47, 919, 1933.
118. **Travassos, J.,** La tique, *Amblyomma striatum*, Koch 1844 comme vecteur de "typhus exanthematique de Sao Paulo." Infection naturelle en specimens recueillis sur des chiens, dans un foyer un foyer do la capitol (Sao Paulo), *C. R. Seances Soc. Biol. Paris*, 127, 1377, 1938.
119. **Wolbach, S. B.,** Studies on Rocky Mountain spotted fever, *J. Med. Res.*, 41, 1, 1919.
120. **Ricketts, H. T.,** The transmission of Rocky Mountain spotted fever by the bite of the wood tick *(Dermacentor occidentalis), JAMA*, 47, 358, 1906.
121. **Ricketts, H. T.,** Further observations on Rocky Mountain spotted fever and *Dermacentor occidentalis, JAMA*, 47, 1067, 1906.
122. **Ricketts, H. T.,** The role of the wood tick (*Dermacentor occidentalis*) in Rocky Mountain spotted fever, and the susceptibility of local animals to this disease, a preliminary report, *JAMA*, 49, 24, 1907.
123. **Ricketts, H. T.,** Further experiments with the wood tick in relation to Rocky Mountain spotted fever, *JAMA*, 49, 1278, 1907.
124. **Ricketts, H. T.,** Observations on the virus and means of transmission of Rocky Mountain spotted fever, *J. Infect. Dis.*, 4, 141, 1907.
125. **Ricketts, H. T.,** A summary of investigations of the nature and means of transmission of Rocky Mountain spotted fever, *Chicago Pathol. Soc. Trans.*, 7, 73, 1907.
126. **Bell, E. J. and Kohls, G. M.,** personal communication.
127. **Parker, R. R.,** Rocky Mountain spotted fever, *JAMA*, 110, 1185, 1938.
128. **Stoenner, H. G., Holdenried, R., Lackman, D., and Orsborn, J. S., Jr.,** The occurrence of *Coxiella burnetii, Brucella* and other pathogens among fauna of the Great Salt Lake Desert in Utah, *Am. J. Trop. Med. Hyg.*, 8, 590, 1959.
129. **Badger, L. F.,** Rocky Mountain spotted fever (eastern type) virus recovered from the dog tick *Dermacentor variabilis* found in nature, *Public Health Rep.*, 47, 2365, 1932.
130. **Brigham, G. D. and Watt, J.,** Highly virulent strains of Rocky Mountain spotted fever virus isolated from ticks (*Dermacentor variabilis*) in Georgia, *Public Health Rep.*, 55, 2125, 1940.

131. **Philip, C. B.,** Rocky Mountain spotted fever; known and potential tick vectors in the United States, *Proc. 6th Pac. Sci. Congr.,* Berkeley, Calif., 5, 581, 1942.
132. **Burgdorfer, W., Cooney, J. C., and Thomas, L. A.,** Zoonotic potential (Rocky Mountain spotted fever and tularemia) in the Tennessee Valley Region. II. Prevalence of *Rickettsia rickettsii* and *Francisella tularensis* in mammals and ticks from land between the lakes, *Am. J. Trop. Med. Hyg.,* 23, 109, 1974.
133. **Burgdorfer, W.,** A review of Rocky Mountain spotted fever (tick-borne typhus), its agent, and its tick vectors in the United States, *J. Med. Entomol.,* 12, 269, 1975.
134. **Burgdorfer, W., Adkins, T. R., and Priester, L. E.,** Rocky Mountain spotted fever (tick-borne typhus) in South Carolina: an educational program and tick/rickettsial survey in 1973 and 1974, *Am. J. Trop. Med. Hyg.,* 24, 866, 1975.
135. **Coultrip, R. L., Emmons, R. W., Legfers, L. F., Marshall, J. D., Jr., and Murray, K. F.,** Survey for the arthropod vectors and mammalian hosts of Rocky Mountain spotted fever and plague at Fort Ord, California, *J. Med. Entomol.,* 10, 303, 1973.
136. **Sexton, D. J., Burgdorfer, W., Thomas, L., and Norment, B. R.,** Rocky Mountain spotted fever in Mississippi: survey for spotted fever antibodies in dogs and for spotted fever group rickettsiae in dog ticks, *Am. J. Epidemiol.,* 103, 192, 1976.
137. **Linneman, C. C., Jr., Schaeffer, A. E., Burgdorfer, W., Hutchinson, L., and Philip, R. N.,** Rocky Mountain spotted fever in Clermont County, Ohio. II. Distribution of population and infected ticks in an endemic area, *Am. J. Epidemiol.,* 111, 31, 1980.
138. **Burgdorfer, W.,** Hemolymph test. A technique for detection of rickettsiae in ticks, *Am. J. Trop. Med. Hyg.,* 19, 1010, 1970.
139. **Burgdorfer, W.,** Tick-borne disease in the United States. Rocky Mountain spotted fever and Colorado tick fever, *Acta Trop.,* 34, 103, 1977.
140. **Burgdorfer, W., Sexton, D. J., Gerloff, R. K., Anacker, R. L., Philip. R. N., and Thomas, L. A.,** *Rhipicephalus sanguineus:* vector of a new spotted fever group rickettsia in the United States, *Infect. Immun.,* 12, 205, 1975.
141. **Parker, R. R.,** Transmission of Rocky Mountain spotted fever by the rabbit tick, *Haemaphysalis leporis-palustris* Packard, *Am. J. Trop. Med.,* 3, 39, 1923.
142. **Parker, R. R., Bell, J. F., Chalgren, W. S., Thrailkill, F. B., and McKee, M. T.,** The recovery of strains of Rocky Mountain spotted fever and tularemia from ticks of the eastern United States, *J. Infect. Dis.,* 91, 231, 1952.
143. **Clifford, C. M., Sonenshine, D. E., Atwood, E. L., Robbins, C. S., and Hughes, L. E.,** Tests on ticks taken from wild birds collected in the eastern United States for rickettsiae and viruses, *Am. J. Trop. Med. Hyg.,* 18, 1057, 1969.
144. **Sonenshine, D. E. and Clifford, C. M.,** Contrasting incidence of Rocky Mountain spotted fever in ticks infecting wild birds in eastern United States piedmont and coastal areas, with notes on the ecology of these ticks, *J. Med. Entomol.,* 10, 497, 1973.
145. **Anigstein, L. and Badger, M. N.,** Investigations on rickettsial diseases in Texas. Part I. Epidemiological role of ticks common to the Gulf Coast in relation to local spotted fever, *Tex. Rep. Biol. Med.,* 1, 105, 1943.
146. **Regendanz, P. and Muniz, J.,** Pasquisas sobrea a transmissao do typho exantematico de Sao Paulo por ixodideos, *Bras. Med.,* 50, 45, 1936.
147. **Silva-Goytia, R. and Elizondo, A.,** Etudos sobre fiebre manchada en Mexico. II. Parasitos hematofagos encontrados naturalmente infectados, *Rev. Med. Mexico,* 32, 278, 1952.
148. **Davis, G. E.,** Experimental transmission of the rickettsiae of the spotted fevers of Brazil, Colombia and the United States by the argasid tick *Ornithodorus nicollei, Public Health Rep.,* 58, 1742, 1943.
149. **Davis, G. E.,** Experimental transmission of the spotted fevers in the United States, Colombia and Brazil by the argasid tick *Ornithodoros parkeri, Public Health Rep.,* 58, 1201. 1943.
150. **Lemos Monteiro da Silva, J., Fonseca, F., and Prado, A.,** *Mem. Inst. Butantan,* 6, 139, 1931; as cited in **Dias, E. and Martins, A. V.,** *Am. J. Trop. Med.,* 19, 103, 1939.
151. **Lemos Monteiro da Silva, J.,** Estudos sobre o typho exanthematico do Sao Paulo, *Mem. Inst. Butantan,* 6, 3, 1931; as cited in **Dias, E. and Martins, A. V.,** *Am. J. Trop. Med.,* 19, 103, 1939.
152. **Davis, G. E.,** The tick *Ornithodoros rudis* as a host to the rickettsiae of the spotted fevers of Colombia, Brazil and the United States, *Public Health Rep.,* 59, 1016,. 1943.
153. **Cooley, R. A.,** The Rocky Mountain Wood tick, Bull. No. 268, *Montana State College Agric. Exp. Stn.,* Bozeman, 1932, 1.
154. **Burgdorfer, W.,** Ecology of tick vectors of American spotted fever, *Bull. W.H.O.,* 40, 375, 1969.
155. **Sonenshine, D. E.,** Zoogeography of the American dog tick *Dermacentor variabilis, Recent Adv. Acarol.,* 2, 123, 1979.
156. **Smith, C. N., Cole, M. M. and Gouch, H. K.,** Biology and control of the American dog tick, *U.S. Dep. Agric. Tech. Bull.,* 905, 1, 1946.

157. **Dias, E. and Martins, A. V.**, Spotted fever in Brazil. A summary, *Am. J. Trop. Med.*, 19, 103, 1939.
158. **Cooley, R. A. and Kohls, G. M.**, The genus *Amblyomma* (Isodidae) in the United States, *J. Parasitol.*, 30, 77, 1944.
159. **Lundgren, D. L., Thorpe, B. D., and Haskell, C. D.**, Infectious diseases in wild animals in Utah. VI. Experimental infection of birds with *Rickettsia rickettsii*, *J. Bacteriol.*, 91, 963, 1966.
160. **Harrel, G. T.**, Rocky Mountain spotted fever, *Medicine*, 28, 333, 1949.
161. **Shepard, C. C. and Topping, N. H.**, Rocky Mountain spotted fever. A study of complement fixation in the serum of certain dogs, *J. Infect. Dis.*, 78, 63, 1946.
162. **Miller, J. K.**, Rocky Mountain spotted fever on Long Island, *Ann. Intern. Med.*, 33, 1398, 1950.
163. **Lechleitner, R. R.**, Some parasites and infectious diseases in a black-tailed jack rabbit population in the Sacramento Valley, California, *Calif. Fish Game*, 45, 83, 1959.
164. **Pagan, F. F., McMahon, K. J., and Bowen, R. E.**, Complement fixing antibodies for *Rickettsia rickettsii* in serums of black-tailed jack rabbits, *Public Health Rep.*, 76, 1120, 1961.
165. **Burgdorfer, W., Newhouse, V. F., Pickens, E. G., and Lackman, D. B.**, Ecology of Rocky Mountain spotted fever in western Montana. I. Isolation of *Rickettsia rickettsii* from wild mammals, *Am. J. Hyg.*, 76, 293, 1962.
166. **Vest, E. D., Lundgren, D. L., Parker, D. D., Johnson, D. E., Morse, E. L., Bushman, J. B., Sidwell, R. W., and Thorpe, B. D.**, Results of a 5 year survey for certain enzootic diseases in the fauna of western Utah, *Am. J. Trop. Med. Hyg.*, 14, 124, 1965.
167. **Gould, D. J. and Miesse, M. L.**, Recovery of a rickettsia of the spotted fever group from *Microtus pennsylvanicus* from Virginia, *Proc. Soc. Exp. Biol. Med.*, 85, 558, 1954.
168. **Shirai, A., Bozeman, F., Perri, S., Humphries, J. W., and Fuller, H. S.**, Ecology of Rocky Mountain spotted fever. I. *Rickettsia rickettsii* recovered from a cottontail rabbit from Virginia, *Proc. Soc. Exp. Biol. Med.*, 107, 211, 1961.
169. **Bozeman, F. M., Shirai, A., Humphries, J. W., and Fuller, H. S.**, Ecology of Rocky Mountain spotted fever. II. Natural infection of wild mammals and birds in Virginia and Maryland, *Am. J. Trop. Med. Hyg.*, 16, 48, 1967.
170. **Burgdorfer, W., Friedhoff, K. T., and Lancaster, J. L.**, Natural history of tick-borne spotted fever in the U.S.A. Susceptibility of small mammals to virulent *Rickettsia rickettsii*, *Bull. W.H.O.*, 35, 149, 1966.
171. **Lundgren, D. L., Nicholes, P. S., and Thorpe, B. D.**, Experimental infection of lagomorphs with *Rickettsia ricketsii*, *Am. J. Trop. Med. Hyg.*, 17, 213, 1968.
172. **Bequaert, J. C.**, The ticks, or Ixodoidea, of northeastern United States and eastern Canada, *Entomol. Am.*, 25, 73, 1946.
173. **Kohls, G. M.**, Vectors of rickettsial diseases, in *The Rickettsial Diseases of Man*, Moulton, F. R., Ed., American Association for the Advancement of Science, Washington, D.C., 1948, 83.
174. **Jellison, W. L.**, The geographical distribution of Rocky Mountain spotted fever and Nuttall's cottontail in the western United States, *Public Health Rep.*, 60, 958, 1945.
175. **Burgdorfer, W., Cooney, J. C., Mavros, A. J., Jellison, W. L., and Maser, C.**, The role of cottontail rabbits (*Sylvilagus* spp.) in the ecology of *Rickettsia rickettsii* in the United States, *Am. J. Trop. Med. Hyg.*, 29, 686, 1980.
176. **Jellison, W. L.**, Rocky Mountain spotted fever; a new look at an old problem, *J. Med. Entomol.*, 9, 595, 1972.
177. **Nelson, E. W.**, The rabbits of North America, North American Fauna, U.S. Department of Agriculture Biological Survey, No. 29, 1909, 1.
178. **Bohlin, B.**, Some mammalian remains from Shi-ehr-ma-Ch'eng, Hui-hui-p'u area, western Kansu, Publ. 35, Rep. Sino-Swedish Sci. Exp. N.W. Provinces China, 1951, 1.
179. **Wood, A. E.**, Eocene radiation and phylogeny of the rodents, *Evolution*, 13, 354, 1959.
180. **Simpson, G. G.**, Holarctic mammalian faunas and continental relationships during the Cenozoic, *Geol. Soc. Am. Bull.*, 58, 613, 1947.
181. **Wood, A. E.**, What, if anything, is a rabbit, *Evolution*, 11, 417, 1957.
182. **Simpson, G. G.**, The principles of classification and a classification of mammals, *Am. Mus. Nat. Hist. Bull.*, 85, 1, 1945.
183. **Wilson, R. W.**, Evolution of the early Tertiary rodents, *Evolution*, 5, 207, 1951.
184. **Parker, R. R., Kohls, G. M., Cox, W., and Davis, G. E.**, Observations on an infectious agent from *Amblyomma maculatum*, *Public Health Rep.*, 54, 1482, 1939.
185. **Parker, R. R.**, A pathogenic rickettsia from the gulf coast tick, *Amblyomma maculatum*, in *Proc. 3rd Int. Congr. Microbiol.*, Dawson, M. H., Ed., International Association of Microbiologists, New York, 1940, 390.
186. **Price, W.**, Interference phenomenon in animal infections with rickettsiae of Rocky Mountain spotted fever, *Proc. Soc. Exp. Biol. Med.*, 82, 180, 1953.

187. **Travassos, J.,** Resultats des inoculations experimental series du virus du "typhus exanthematique de Sao Paulo" a la sarique marsupiale *(Didelphis aurita), C. R. Seances Soc. Biol. Paris,* 126, 1049, 1937.
188. **Price, W. H.,** Variations in virulence of *Rickettsia rickettsii* under natural and experimental conditions, in *Dynamics of Viral and Rickettsial Infections,* Blakiston, Philadelphia, 1954, 164.
189. **Bell, E. J., Kohls, G. M., Stoenner, H. G., and Lackman, D. B.,** Nonpathogenic rickettsias related to the spotted fever group isolated from ticks, *Dermacentor variabilis* and *D. andersoni* from eastern Montana, *J. Immunol.,* 90, 770, 1963.
190. **Feng, W. C., Murray, E. S., Burgdorfer, W., Spielman, J. M., Rosenberg, G., Dang, K., Smith, C., Spickert, C., and Waner, J. L.,** Spotted fever group rickettsiae in *Dermacentor variabilis* from Cape Cod, Massachusetts, *Am. J. Trop. Med. Hyg.,* 29, 691, 1980.
191. **Feng, W. C. and Waner, J. L.,** Serological cross reactions and cross protection in guinea pigs infected with *Rickettsia rickettsii* and *Rickettsia montana, Infect. Immun.,* 28, 627, 1980.
192. **Weyer, F.,** Rickettsiae and lice, in *2nd Int. Symp. Rickettsiae Rickettsial Dis.,* Kazar, J., Ormsbee, R. A., and Tarasevich, I. N., Eds., Slovak Academy of Sciences, Bratislava, Czechoslovakia, 1978, 515.
193. **Cox, H. R.,** Cultivation of the rickettsiae of the Rocky Mountain spotted fever, typhus and Q-fever groups in the embryonic tissues of developing chicks, *Science,* 94, 399, 1941.
194. **Ormsbee, R., Peacock, M., Gerloff, R., Tallent, G., and Wike, D.,** Limits of rickettsial infectivity, *Infect. Immun.,* 19, 239, 1978.
195. **Baltazard, M., Bahmanyar, M., Mofidi, C., and Seydian, B.,** Le foyer de peste du Kurdistan, *Bull. W.H.O.,* 5, 441, 1952.
196. **Baltazard, M., Seydian, B., Mofidi, C., Bahmanyar, M., and Pournaki, R.,** The resistance of certain species of wild rodents to plague; observations made in nature, *Ann. Inst. Pasteur,* 85, 411, 1953.
197. **Baltazard, M., Bahmanyar, M., Mostachfi, P., Eftekhari, M., and Mofidi, C.,** Recherches sur la peste en Iran, *Bull. W.H.O.,* 23, 141, 1960.
198. **Quan, S. F. and Kartman, L.,** Ecological studies of wild rodent plague in the San Francisco Bay Area of California. VIII. Susceptibility of wild rodents to experimental plague infection, *Zoonoses Res.,* 1, 121, 1962.
199. **Marchette, N. J.,** *Pasteurella pestis* of low virulence for mice and guinea pigs isolated from a *Peromyscus* plague focus in Utah, *Am. J. Trop. Med. Hyg.,* 12, 215, 1963.
200. **Baker, A. C.,** The typical epidemic series, *Am. J. Trop. Med.,* 23, 559, 1943.
201. **Anigstein, L. and Badger, M. N.,** Investigations on rickettsial diseases in Texas. Part I. Epidemiological role of ticks common to the Gulf Coast in relation to local spotted fever, *Tex. Rep. Biol. Med.,* 1, 105, 1943.
202. **Hayes, S. F. and Burgdorfer, W.,** Ultrastructure of *Rickettsia rhipicephali,* a new member of the spotted fever group rickettsiae in tissues of the host vector, *Rhipicephalus sanguineus, J. Bacteriol.,* 137, 605, 1979.
203. **Burgdorfer, W., Brinton, L. P., Krynski, W. L., and Philip, R. N.,** *Rickettsia rhipicephali,* a new spotted fever group rickettsia from the brown dog tick, *Rhipicephalus sanguineus,* in *2nd Int. Symp. Rickettsiae Rickettsial Dis.,* Kazar, J., Ormsbee, R. A., and Tarasevich, I. N., Eds., Slovak Academy of Sciences, Bratislava, Czechoslovakia, 1978, 307.
204. **Plotz, H., Reagan, R. L., and Wertman, K.,** Differentiation between fievre boutonneuse and Rocky Mountain spotted fever by means of complement fixation, *Proc. Soc. Exp. Biol. Med.,* 55, 173, 1944.
205. **Sussman, L. N.,** Kew Gardens' spotted fever, *N.Y. Med.,* 2, 27, 1946.
206. **Shankman, B.,** Report on an outbreak of endemic febrile illness, not yet identified, occurring in New York City, *N.Y. State J. Med.,* 46, 2156, 1946.
207. **Roueche, B.,** The alerting of Mr. Pomerantz, *New Yorker,* 28, August 30, 1947.
208. **Greenberg, M., Pelliteri, O., Klein, I. S., and Huebner, R. J.,** Rickettsialpox—a newly recognized rickettsial disease. II. Clinical findings, *JAMA,* 133, 901, 1947.
209. **Huebner, R. J., Stamps, P., and Armstrong, C.,** Rickettsialpox—a newly recognized rickettsial disease. I. Isolation of the etiological agent, *Public Health Rep.,* 61, 1605, 1946.
210. **Huebner, R. J., Jellison, W. I., and Armstrong, C.,** Rickettsialpox—a newly recognized rickettsial disease. V. Recovery of *Rickettsia akari* from a house mouse *(Mus musculus). Public Health Rep.,* 62, 777, 1947.
211. **Lackman, D. B.,** A review of information on rickettsialpox in the United States, *Clin. Pediatr.,* 2, 296, 1963.
212. **Shepard, C. C., Redus, M. A., Tzianabos, T., and Warfield, D. T.,** Recent experience with the complement fixation test in the laboratory diagnosis of rickettsial diseases in the United States, *J. Clin. Microbiol.,* 4, 277, 1976.

213. **Wong, B., Singer, C., Armstrong, D., and Millian, S. J.,** Rickettsialpox. Case report and epidemiological review, *JAMA*, 242, 1998, 1979.

214. **Tarasevic, I. V.,** Natural Focality and Epidemiology of Endemic Rickettsioses, W.H.O. Inter-regional Traveling Semin. Natural Foci Zoonoses, Moscow, 1974, 1.

215. **Brezina, R., Urvolgyi, J., Rosicky, B., Cilka, S., Dushniku, N., Naracik, K., and Dishnica, G.,** Rickettsioses and infections caused by viruses of the psittacosis-ornithosis-mammalian pneumonia group in Albania, *J. Hyg. Epidemiol. Microbiol. Immunol.*, 5, 85, 1961.

216. **Terzin, A. L. and Gaon, J.,** Some viral and rickettsial infections in Bosnia and Herzegovina. A sero-epidemiological study, *Bull. W.H.O.*, 15, 299, 1956.

217. **Rehacek, J.,** Spotted fever group rickettsiae in Europe, *Recent Adv. Acarol.*, 2, 245, 1979.

218. **Hirst, S.,** On the parasitic acari found on the species of rodents frequenting human habitation in Egypt, *Bull. Entomol. Res.*, 5(3), 215, 1914.

219. **Pratt, H. D., Lane, J. E., and Harmiton, F. C.,** New locality records for *Allodermanyssus sanguineus*, vector of rickettsialpox, *J. Econ. Entomol.*, 42, 414, 1949.

220. **Philip, C. B. and Hughes, L. E.,** The tropical rat mite, *Liponyssus bacoti*, as an experimental vector of rickettsialpox, *Am. J. Trop. Med.*, 28, 697, 1948.

221. **Furman, D. P. and Radovsky, F. J.,** A new species of *Ornithonyssus* from the white-tailed antelope squirrel, with a rediagnosis of the genus *Ornithonyssus*, *Pan Pac. Entomol.*, 39, 75, 1963.

222. **Krantz, G. W.,** New synonymy in the Dermanyssinae Kolenati, 1859, with a description of a new species of *Dermanyssus* (Acarine, Dermanyssidae), *Proc. Entomol. Soc. Wash.*, 61, 174, 1959.

223. **Rose, R. M.,** The clinical manifestations and laboratory diagnosis of rickettsialpox, *Ann. Intern. Med.*, 31, 871, 1949.

224. **Pedersen, C. E., Jr. and Walters, V. D.,** Comparative electrophoresis of spotted fever group rickettsial proteins, *Life Sci.*, 22, 583, 1978.

225. **Ormsbee, R., Peacock, M., Philip, R., Casper, E., Plorde J., Gabrehidan, T., and Wright, L.,** Antigenic relationships between typhus and spotted fever groups of rickettsiae, *Am. J. Epidemiol.*, 108, 53, 1978.

226. **Andrew, R., Bonnin, J. M., and Williams, S.,** Tick typhus in North Queensland, *Med. J. Aust.*, 2, 253, 1946.

227. **Funder, J. F. and Jackson, A. V.,** North Queensland tick typhus: a comparative study of the rickettsia with that of murine typhus, *Med. J. Aust.*, 2, 258, 1946.

228. **Campbell, R. W. and Pope, J. H.,** The value of newborn mice as a sensitive host for *Rickettsia australis*, *Aust. J. Sci.*, 30, 324, 1968.

229. **Philip, C. B.,** Miscellaneous human rickettsioses, in *Communicable Diseases*, Pullen, R. L., Ed., Lea and Febiger, Philadelphia, 1950, 781.

230. **Knyvett, A. F. and Sandars, D. F.,** North Queensland tick typhus: a case report defining a new endemic area, *Med. J. Aust.*, 2, 592, 1964.

231. **Domrow, R. and Derrick, E. H.,** *Ixodes holocyclus*, the man-biting tick in southeast Queensland, *Aust. J. Sci.*, 27, 234, 1965.

232. **Campbell, R. W. and Domrow, R.,** Rickettsiosis in Australia: Isolation of *Rickettsia tsutsugamushi* and *Rickettsia australis* from naturally infected arthropods, *Trans. R. Soc. Trop. Med. Hyg.*, 68, 397, 1974.

233. **Roberts, F. H. S.,** A systematic study of the Australian species of the genus *Ixodes*, *Aust. J. Zool.*, 8, 392, 1960.

234. **Cook, I. and Campbell, R. W.,** Rickettsiosis—North Queensland tick typhus, *Rep. Queens. Inst. Med. Res.*, 20, 4, 1965.

235. **Campbell, R. W. and Domrow, R.,** Rickettsioses in Australia: Ecology of *Rickettsia tsutsugamushi* and *Rickettsia australis*, in *2nd Int. Symp. Rickettsiae Rickettsial Dis.*, Kazar, J., Ormsbee, R. A., and Tarasevich, I. N., Eds., Slovak Academy of Sciences, Bratislava, Czechoslovakia, 1978, 505.

236. **Kouwenaar, W. and Wolff, J. W.,** Onderzoehingen over sumatraansche Rickettsiosen, *Geneeskd. Tijdsch. Neder. Indie*, 74, 1659, 1934.

237. **Hughes, L. E., Clifford, C. M., Gresbrink, R., Thomas, L. A., and Keirans, J. E.,** Isolation of a spotted fever group rickettsia from the Pacific Coast tick *Ixodes pacificus* in Oregon, *Am. J. Trop. Med. Hyg.*, 25, 513, 1976.

238. **Cooley, R. A. and Kohls, G. M.,** The genus *Ixodes* in North America, No. 184, U.S. Natl. Inst. Health Bull., Washington D.C., 1945, 1.

239. **Philip, C. B. and Hughes, L. E.,** Disease agents found in the rabbit tick, *Dermacentor parumapertus*, in the southwestern United States, *Abstr. 6th Int. Congr. Microbiol.*, International Congress for Microbiology, Rome, 1953, 541.

240. **Cory, J., Yunker, C. E., Howarth, J. A., Hokama, Y., Hughes, L. E., Thomas, L. A., and Clifford, C. M.,** Isolation of spotted fever group and *Wolbachia* like agents from field collected materials by means of plaque formation in mammalian and mosquito cells, *Acta Virol. Engl. Ed.*, 19, 443, 1975.
241. **Parker, R. R., Philip, C. B., Davis, G. E., and Cooley, R. A.,** Ticks of the United States in relation to disease in man, *J. Econ. Entomol.*, 30, 51, 1937.
242. **Kohls, G. M.,** Hosts of the immature stages of the Pacific Coast tick *Dermacentor occidentalis* Neum. (Ixodidae), *Public Health Rep.*, 52, 490, 1937.
243. **Woodland, J. C., McDowell, M. M., and Richards, J. T.,** Bullis fever (Lone Star fever—tick fever) an endemic disease observed at Brooke General Hospital, Fort Sam Houston, Texas, *JAMA*, 122, 1156, 1943.
244. **Anigstein, L. and Anigstein, D.,** A review of the evidence in retrospect for a rickettsial etiology in Bullis fever, *Tex. Rep. Biol. Med.*, 33, 201, 1975.
245. **Livesay, H. R. and Pollard, M.,** Laboratory report on a clinical syndrome referred to as Bullis fever, *Am. J. Trop. Med.*, 23, 475, 1943.
246. **Livesay, H. R. and Pollard, M.,** Serological studies of Bullis fever, *Am. J. Trop. Med.*, 24, 281, 1944.
247. **Anigstein, L. and Bader, M. N.,** Investigations on rickettsial diseases in Texas. Experimental study of Bullis fever, *Tex. Rep. Biol. Med.*, 1, 389, 1943.
248. **Bader, M. N. and Anigstein, L.,** Specificity of Bullis fever rickettsia, *Tex. Rep. Biol. Med.*, 2, 405, 1944.
249. **Pollard, M., Livesay, H. K., Wilson, D. J., and Woodland, J. C.,** Experimental studies with Bullis fever, *Am. J. Trop. Med.*, 26, 175, 1946.
250. **Calhoun, E. L.,** Natural occurrences of tularemia in the lone star tick, *Amblyomma americanum* (L.) and in dogs in Arkansas, *Am. J. Trop. Med. Hyg.*, 3, 360, 1954.
251. **Kokernot, R. H., Calisher, C. H., Stannard, L. J., and Hayes, J.,** Arbovirus studies in the Ohio-Mississippi Basin, 1964–1967. VII. Lone Star virus, a hitherto unknown agent isolated from the tick *Amblyomma americanum* (Linn.), *Am. J. Trop. Med. Hyg.*, 18, 789, 1969.
252. **Roberts, F. H. S.,** Further observations on the Australian species of *Aponomma* and *Amblyomma* with descriptions of the nymphs of *Amblyomma moreliae* (L. Koch) and *Amblyomma loculosum* Neumann (Acarina: Ixodidae), *Aust. J. Zool.*, 12, 288, 1964.

Chapter 6

THE TYPHUS COMPLEX: *RICKETTSIA TYPHI* AND *R. PROWAZEKII* ADAPTATION TO INSECTS

N. Marchette

TABLE OF CONTENTS

I. Characteristics of the Typhus Group 114
- A. Antigenic Characteristics 114
- B. Molecular Biology 114

II. Geographical Distribution and Host Range 114
- A. *Rickettsia typhi* 114
- B. *Rickettsia canada* 116
- C. *Rickettsia prowazekii* 118
  1. Introduction 118
  2. Present Distribution 118
  3. Separation of *R. typhi* from *R. prowazekii* 118
  4. Presence in Ticks and Domestic Animals 119
     - a. Early Serological Evidence in Africa 119
     - b. Studies in Ethiopia and Egypt 120
  5. Antibody in Wild Vertebrates 121
  6. Experimental Infection of Domestic Animals 122
  7. Experimental Infection of Ticks 123
  8. Hypotheses Concerning Origin of Typhus Antibodies in Domestic Animals 124
  9. Maintenance of *R. prowazekii* in Man 125

III. Origin of Typhus Rickettsiae 126
- A. Early Ideas on the Relationship of *R. prowazekii* to *R. typhi* 126
- B. Evolution of *Rickettsia typhi* 127
  1. Early Thoughts on Place of Origin 127
  2. Mites as Original Hosts 128
  3. *Xenopsylla cheopis* 129
  4. The Rat 130
- C. Evolution of *Rickettsia prowazekii* 131
  1. Derivation from *R. typhi*—Early Ideas 131
  2. Origin and Dispersion of *Rattus rattus* 132
     - a. Migrations of *Rattus* to Europe and Beyond 133
     - b. The History of Plague in the Middle East and Europe 135
  3. History of Early Typhus Epidemics 137
  4. Origin and Dispersion of *Rattus norvegicus* 139
  5. Flying Squirrels 140
  6. The Human Body Louse 141
  7. Typhus—A Man-Made Malady? 142

IV. Summary 143

References 144

I. CHARACTERISTICS OF THE TYPHUS GROUP

The typhus complex classically contained only the two species of the Rickettsieae transmitted by insects, but now may include a third member associated with ticks. *Rickettsia typhi,* the etiologic agent of fleaborne (murine or endemic) typhus, is a natural parasite (symbiote) of *Rattus* spp. and their ectoparasitic fleas, *Xenopsylla cheopis,* and lice, *Polyplax spinulosus. Rickettsia prowazekii,* the etiologic agent of louseborne (epidemic or classical) typhus, until recently was known to naturally infect only the human body louse, *Pediculus humanus,* and man. A possible third member of the typhus complex, *R. canada,* naturally infecting ticks in North America was discovered in 1967.[1] The taxonomic position of *R. canada* is not entirely clear. It appears to be more closely related biologically to the tick-typhus group than to either fleaborne or louseborne typhus, but antigenically it belongs in the typhus complex.

A. Antigenic Characteristics

The typhus rickettsiae possess a common soluble group antigen, detectable by the complement fixation (CF) test, which separates them from the tick-typhus or spotted fever group rickettsiae. It is clear, however, that antigenic determinants are shared to some extent with spotted fever organisms.[2-6] Each of the typhus rickettsiae also possess cell-associated type-specific antigens demonstrable in the CF test using washed rickettsiae as antigens, the toxin neutralization test, and by using type-specific mouse antisera. Cross-protection tests in guinea pigs further distinguish tick typhus from typhus group rickettsiae and, since the protection is only partial, demonstrates a relationship between them. The recently developed microimmunofluorescence (MIF) test also detects antigenic differences between typhus and spotted fever group rickettsiae and clearly distinguishes each of the three typhus organisms from each other.[8]

Antigenically and biologically *R. typhi* is more closely related to *R. prowazekii* than to *R. canada.* In the MIF test, *R. canada* is antigenically unique and not at all closely related to either *R. typhi* or *R. prowazekii.*[8] Other serological data suggest that *R. canada* shares surface antigens with spotted fever rickettsiae and internal antigens with *R. prowazekii.*[9-11] All three species replicate in the human body louse, but *R. canada* is not pathogenic for the louse and replicates only in cells of the intestinal epithelium.[12] All other rickettsiae pathogenic for man replicate extracellularly in the lumen of the louse gut.[13]

B. Molecular Biology

On the molecular level, the DNA of the typhus rickettsiae and *R. canada* contain identical percentages of guanine + cytosine. The proportion of these two bases is slightly lower than in spotted fever group DNA.[14-16] Analysis of the proteins of purified rickettsiae by SDS acrylamide gel electrophoresis also has shown distinct differences in number of proteins and their molecular weights between the typhus organisms and three spotted fever organisms (*R. rickettsii, R. conorii,* and *R. akari*).[17-19]

The close relationship between *R. prowazekii* and *R. typhi* has been demonstrated in DNA hybridization studies, which consistently show 70 to 77% homology among several strains of each species.[15] This confirms the work done by others and by Dasch et al.,[20] who showed, in a variety of biochemical analyses, that *R. prowazekii* and *R. typhi* are distinct species more closely related to each other than either is to *R. canada.*

II. GEOGRAPHICAL DISTRIBUTION AND HOST RANGE

A. *Rickettsia typhi*

Fleaborne typhus is distributed throughout most of the world[21] as a natural infection of domiciliated rats and mice. It is presumably transmitted from rat to rat by the rat louse, *Polyplax spinulosus,*[22] and by the oriental rat flea, *Xenopsylla cheopis.*[23–25] A wide variety of rodents and other small domestic and wild animals are susceptible to experimental infection,[26,27] but it is not known what role they play in the ecology of *R. typhi*. Complement–fixing antibody to *R. typhi* has been demonstrated in many small mammals in the southern U.S.[28] Strains of *R. typhi* have been isolated from *Peromyscus polionotus* in Georgia,[29] *Mus musculus* (house mice) in Tunis,[30] and *Mus wagneri* (and from mouse fleas, *Leptopsylla musculi*) in Peking.[31] Ormsbee et al.[32] found serological evidence of natural fleaborne typhus infection in *Arvicanthis* (species not given) in Egypt. The house shrew, *Suncus murinus,* was shown to be naturally infected in Malaya,[33,34] and the organism has been recovered from *Rattus exulans* as well as from *R. rattus* in Thailand.[35] In India, serological evidence of infection and presumptive isolation of *R. typhi* have been reported for three species of house mice (*M. booduga, M. platyhrix, M. saxicola*), *Rattus rattus satarae, R. blandfordi, Golunda ellioti* (bush rat), *Milardia kondara* (soft-furred field rat), and *S. murinus.*[35a]

In East Africa, *R. typhi* has been isolated only from *Rattus rattus,* which are the maintaining hosts,[36] but complement-fixing antibody has been found in sera of a few wild rodents—*Arvicanthis niloticus, Mastomys natalensis, Lophuromys flavopunctatus, Tatera robusta,* and *Otomys* sp.[37–39]

Kalra and Rao[40] reported the isolation of *R. typhi* from cattle ticks (*Boophilus microplus = B. australis*) in India, but the significance of this is unknown. The boophilids are one-host ticks, mainly on cattle, and are host specific. This species was reportedly common in the game parks and in fields around Jubulpore and is known to occur on wild animals and birds.[41] Reiss-Gutfreund[42] also reported the isolation of *R. typhi* from cattle ticks (*Hyalomma* sp. nr. *truncatus*) near Addis Ababa, Ethiopia. There are reports of isolation of *R. typhi* from the mesostigmatic mite, *Ornythonyssus bacoti,* collected from house rats in China.[43,44] In these instances, the rats were also infected and the rickettsiae were probably present in the mite's blood meal.

Fleaborne typhus was called urban or "shop" typhus in Malaya to distinguish it from rural (chiggerborne) typhus,[45] and the name of the disease in man may be still appropriate in some places.[46,47] The only well-described natural cycle involves *Rattus rattus* and perhaps one or two other highly domiciliated animals. Occurrence of *R. typhi* infection in wild rodents may be the result of secondary exposure to fleas or lice from urban rats. Natural cycles in wild animals other than in *Rattus* cannot be ruled out entirely, although *R. typhi* seems to be equally as domiciliated as its major vertebrate hosts.

Most microbiology textbooks positively state that murine typhus is transmitted from rat to rat by the rat louse, *Polyplax spinulosus* and the oriental rat flea, *Xenopsylla cheopis*. The paper usually referred to for proof of transmission of *R. typhi* by *P. spinulosus* is that of Mooser et al.[22] The one experiment they described, and from which the conclusions are drawn, was not properly controlled. In it they demonstrated (1) the presence of rickettsiae in lice after they had fed on a rickettsemic rat, and (2) antibody conversion in a single healthy rat after being kept in a jar for some days with a louse-infested rat that had been inoculated with *R. typhi*. This is by no means proof of biological transmission by the louse since replication of the organism in the louse was not demonstrated, nor was the possibility of direct rat-to-rat transmission through infectious urine ruled out.

The paper often cited as proof that *Xenopsylla cheopis* can transmit *R. typhi*[24] is subject to the same criticism. Although these experiments were somewhat more elaborate than those of Mooser et al.,[22] they showed only that uninfected rats placed in a container with infected rats and fleas become infected. In a later paper, Dyer et al.[23] reported much more sophisticated (and controlled) experiments that do strongly suggest that *Xenopsylla cheopis* can transmit *R. typhi* to uninfected rats. This is the reference that should be cited. The closeness of the relationship between *X. cheopis* and *R. typhi* may be illustrated by the observation that the rickettsia lies directly in the cytoplasm of mid-gut epithelial cells, as it does in mammalian cells.[48]

Experimental transmission among rodents has been claimed (but not proven) for *Xenopsylla astia,*[49] *Pulex irritans,*[50] and *Ornithonyssus bacoti.*[51,52] In all these cases, transmission is probably through contamination of the bite wound with infectious feces from the engorging ectoparasite.[53] Violle[54] suggested that rats can cause infection by contamination of food with their own urine or flea feces from their fur, but Worth and Rickard[55] and Rickard[56] were unable to demonstrate transmission experimentally by these routes. The presence of rickettsiae in the urine of infected rats has been reported,[57,58] but it is not known how long the rickettsiae are shed. The experiments of Worth and Rickard[55] suggest that *R. typhi* may not remain viable for very long in rat urine.

The implication of cat fleas, *Ctenocephalus felis,* in human fleaborne typhus infection[59] is pure speculation based on only the most tenuous epidemiologic association. The evidence that the stick-tight flea, *Echidnophaga gallinacea,* plays any role in *R. typhi* ecology is equally untenable. Brigham[60] isolated *R. typhi* from pools of this flea collected from rats on a chicken farm. The rats and their fleas, *X. cheopis,* were also infected. Alicata's[61] claim of experimental transmission of *R. typhi* by *E. gallinacea* is also unconvincing. The experiments showed merely that the fleas could pick up the rickettsiae from infected rats, excrete it in their feces and produce typhus when inoculated into guinea pigs. No natural transmissions were attempted, although fecal excretion of the organism would suggest that they are at least temporarily infectious. Evidence for replication of the rickettsiae in the louse was not given.

The importance of the rat nest as a microfocus of infection has not been fully realized. Both the rat flea and the rat mite breed in the nest and spend part of their lives there. It seems highly probable that *R. typhi* can persist in infectious feces in the nest. If so, then it is reasonable to assume that infant rats may become infected even before they leave the nest and that they harbor the organism in a latent form for long periods, if not for life. Periodic shedding of rickettsiae into the blood stream would provide an intermittent source of organisms for ingestion by fleas, lice, and mites.

The flea is the most efficient vector of the three common ectoparasites of domiciliated rodents, but this may be partly a reflection of its size and the quantity of infectious feces it excretes. It may actually be an amplifying host as well as a transmitter of the agent indirectly through its feces. The louse, which leaves its host only when the rat dies, would seem to be a rather poor vector for continuous maintenance of the cycle, and rat mites are very inefficient vectors experimentally.

B. *Rickettsia canada*

Rickettsia canada is the most recently described typhus group organism. The two existing strains were both recovered from *Haemaphysalis leporispalustris* ticks collected in Ontario, Canada. One positive tick pool was obtained from sentinel domestic rabbits placed in the field; the other from a wild snowshoe hare (*Lepus americanus*) shot in the same general area.[1] Clifford et al.[62] reported serologic evidence suggesting the presence of *R. canada* in *H. leporispalustris* collected from a veery in the eastern U.S. More recently, in a detailed serologic analysis of 70 cases of Rocky Mountain spotted fever (diagnosed clinically or serologically), Bozeman et al.[62] made a presumptive diagnosis of infection with *R. canada* in four cases on the basis of complement fixation tests. One of the patients was from Texas; the other three were from North Carolina. An *R. canada*-like organism has been isolated recently from a *Haemaphysalis leporispalustris* tick removed from a black-tailed jack rabbit (*Lepus californicus*) in

California.[64] There is no definitive evidence that *R. canada* occurs outside North America, but Hamidi et al.[65] reported the presence of complement-fixing antibodies in wild rodents in eastern Iran. Unfortunately, the specificity of the test was not given and the data are difficult to evaluate.

There is not a great deal of evidence for the occurrence of typhus rickettsiae other than *R. canada* in acarine hosts. Kalra and Rao[40] reported *R. typhi* from *Boophilus australus* (=*B. microplus*) and Heisch et al.[39] found evidence of infection in *Rhipicephalus simus* and *Amblyomma variegatum*, but the organism could easily have been picked up incidentally by these ticks while feeding on infected domiciliated rodents. In India, the same species of rat (often the same individuals) may be found in houses (and hence be flea-infested) and in the surrounding open country (and hence be tick-infested). Evidence does not suggest that these isolates were other than *R. typhi* as reported; on the other hand, techniques were not then available for distinguishing between *R. typhi* and *R. canada*. *Rickettsia typhi* has even been reported from the trombiculid mite, *Ascoschongastia indica*,[66] a very common parasite of house rats in Southeast Asia. The significance of such isolations to the ecology of the organism is questionable.

The only other evidence of *R. typhi* in wild-animal ticks about which I am aware is the report of a presumptive isolation from a mixed batch of ticks collected from various hosts in West Malaysia.[33] The rickettsia may have come from *Amblyomma javanense* collected from a pangolin (*Manis javanicus*), but attempts to reisolate it from the portion of the original pool containing only *A. javanense* were unsuccessful. The other ticks in the original mixed pool also failed to elicit illness or antibody response in guinea pigs when reinoculated separately.

Coming as it does on the heels of reports by Reiss-Gutfreund[42,67,68] on the isolation of *R. prowazekii* and *R. typhi* from domestic animal ticks, the recovery of *R. canada* from rabbit ticks and the isolation of *R. prowazekii* from squirrels[69,70] suggests that we may not know all we thought we did about the ecology of louseborne and fleaborne typhus. These developments require, at the very least, a reassessment of basic cycles of insectborne typhus rickettsiae. The new data are not too surprising, however, since they more or less directly support the generally well-accepted hypothesis of the ultimate origin of *R. typhi* and *R. prowazekii* from tickborne ancestors.

On the other hand, we must be cautious about leaping to conclusions on the basis of limited information. Recent experiments by Burgdorfer[71] suggest that birds rather than hares may be the natural vertebrate hosts of *R. canada*. Snowshoe hares experimentally infected with *R. canada* did not develop a detectable rickettsemia, whereas birds (chickens and pheasants) circulated rickettsiae in high titer in the blood. It should be remembered that *H. leporispalustris* commonly parasitizes birds as well as hares. Meadow mice, *Microtus pennsylvanicus*, also developed rickettsemias sufficient to infect *H. leporispalustris* and *D. andersoni*. *Rickettsia canada* has not been recovered from *D. andersoni* in nature, but transovarial transmission was demonstrated experimentally in this tick and not in *H. leporispalustris*.[72]

Ignatovich and Grokhovskaya[73] studied experimental infection of *R. canada* in the argasid ticks *Ornithodoros papillipes* and *O. moubata* and the ixodid tick *Hyalomma dromedarii*. After intracoelomic inoculation, the rickettsia persisted for up to two years in the argasids. Infection could also be acquired by feeding on infected guinea pigs, but no transmission to normal animals could be demonstrated. Transovarial transmission was reported in *O. papillipes* and *H. dromedarii*, but the relevance of these experiments to the ecology and origin of *R. canada* is not clear.

This is an extremely interesting organism since it is clearly related antigenically to the insectborne typhus group rather than to the tickborne typhus group[20] even though it appears to be tickborne. Unlike *R. prowazekii* and *R. typhi,* it also shares with tick-typhus organisms the capacity for growth in the nuclei of infected cells.[74] It may be premature to place *R. canada* in either the typhus or the spotted-fever group. It is conceivable that this organism is a member of a phylogenetically distinct group of yet to be discovered rickettsiae.

C. *Rickettsia prowazekii*

1. *Introduction*

Although louseborne (epidemic) typhus must rank with bubonic plague, smallpox, and typhoid fever as one of the great scourges of mankind, it wasn't until early in the present century that Nicolle and colleagues[75,76] demonstrated that the disease is transmitted by the host-specific human body louse, *Pediculus humanus.* Long before, and for reasons not entirely clear but at least partly related to the sanitary revolution, typhus and plague had begun to wane. The last great typhus pandemic occurred during and following the 30 Years' War in the 17th century. Typhus remained a serious problem in Europe, particularly in jails and institutions and accompanying famines and wars, but it gradually ceased to be a disease of pandemic proportions. In the great upheaval of the First World War, typhus again caused serious epidemics in Eastern Europe and Russia, but strict sanitary precautions among the Western Allies prevented it being a serious problem in Western Europe. Epidemics were numerous again during World War II, but the development of vaccines and especially the use of DDT and major delousing campaigns among prisoners and displaced persons repeatedly disrupted the chain of transmission and kept the disease under control. The epidemiology of louseborne typhus was deceptively simple. *Rickettsia prowazekii* was not known to naturally infect any animals except man and the human body louse until Reiss-Gutfreund[42,67,68,77] reported isolations from ticks and domestic animals in Ethiopia, and Bozeman et al.[69] isolated strains from flying squirrels in the eastern U.S. These developments suggest a hitherto unsuspected complexity to *R. prowazekii* ecology.

2. *Present Distribution*

Rickettsia prowazekii, alone among all the rickettsiae, is capable of causing major epidemics. Louseborne typhus is essentially world-wide in distribution,[78,79] occurring wherever there are concentrations of people infested with body lice. It is more prevalent in the temperate zones than in the tropics where it occurs mostly in higher elevations.[21] This is undoubtedly due to the distribution of body lice. Although *R. prowazekii* readily infects head lice, *Pediculus humanus capitus,*[80] which is more widespread than the body louse, *P. h. corporis,* the distribution of epidemic typhus coincides only with that of the body louse. The development of vaccines, the introduction of antibiotic therapy, and application of potent insecticides in delousing campaigns since World War II have eliminated the disease from much of Europe, Asia, North America, and South America. The most important remaining foci are found in Africa with smaller ones in Central and South America.[81]

3. *Separation of R. typhi from R. prowazekii*

Rickettsia prowazekii is closely related to *R. typhi,* and the early workers were greatly handicapped by lack of reliable methods for distinguishing between them and for accurately diagnosing the diseases they caused.[82] Since both forms produce high titered *Proteus* OX 19-agglutinating antibodies in man and some experimental animals, the two diseases could only be readily separated on epidemiological grounds. Fleaborne typhus was usually less severe than louseborne typhus, tended to occur more often in rural than in urban areas, did not cause large epidemics, and was associated with rats and fleas instead of humans and lice. The rickettsial strains isolated were characterized primarily by their pathogenicity for guinea pigs and rats. Monkeys, however, also were used, and gerbils found favor with some workers. Even with the painstaking work of investigators such as Mooser, Nicolle, Zinsser, Dyer, and others, the results of different laboratories were difficult to compare. The species of experimental animals, or at best the strains, often differed from one laboratory to another. Inoculation procedures were not standardized, and often the number of rickettsiae inoculated was either not known or could only be crudely estimated. Natural variation among newly isolated strains was often not taken into consideration and led to many erroneous conclusions. It is to the credit of the typhus pioneers that they were able to distinguish between the two types of typhus as well as they did. The ability of *R. prowazekii* to infect and multiply in *Xenopsylla cheopis* and of *R. typhi* to do so in *Pediculus humanus* only added to the confusion. Moreover, the human louse can also transmit *R. typhi* as well as *R. prowazekii* from man to man under the proper circumstances.

4. Presence in Ticks and Domestic Animals

Rickettsia prowazekii has long been considered to be an exclusive parasite of man and the human body louse, albeit a recently evolved species not yet too well adapted to either its arthropod or vertebrate host. The ecological situation has thus far been favorable enough for it to survive, and the postulated evolutionary history does not require a nonhuman maintaining host at this stage in its history. If *R. prowazekii* evolved directly from *R. typhi,* which in turn evolved millenia ago from a tickborne ancestor, a tick-vertebrate maintaining system would be superfluous or, at most, secondarily derived. *Rickettsia prowazekii* would seem to be removed from any obligate association with ticks. Any rickettsia of the *R. prowazekii–R. typhi* complex existing in a natural tick-vertebrate cycle should resemble *R. typhi* more closely than *R. prowazekii* if it is phylogenetically related to the typhus branch. The recently isolated *R. canada,* serologically more closely related to *R. typhi* than to *R. prowazekii,* may be just such an example, but a phylogenetic relationship has yet to be established. The recent *R. prowazekii* squirrel isolates have not yet been associated with ticks and their transmission cycle is unknown. However, the existence of basic extrahuman cycles of infection can no longer be discounted out of hand.

a. Early Serological Evidence in Africa

Before the discovery of *R. canada* and the squirrel agents, investigators in Africa began publishing studies suggesting the existence of tick-domestic animal-*R. prowazekii* cycles. The first reports of possible infection of domestic animals with *R. prowazekii* are those of Giroud and his colleagues in Africa,[83,85] who based their conclusions on results of serological tests of questionable validity. The serological data were obtained with crude antigens in the complement fixation test and a microagglutination test developed by Giroud and Giroud[86] upon which the French workers have relied heavily. The antigen used is a crude suspension of infected mouse or rabbit lung and apparently gives such a broad reaction that Giroud and his colleagues even began questioning the specificity of their results, especially since they were unable to isolate *R. prowazekii* from any of the domestic animals in the sero-positive regions. Giroud and Le Gac[87] had earlier claimed that their *R. prowazekii* antigen cross-reacted with antibody from boutonneuse fever patients. This would not be surprising, but their sparse data are woefully inadequate to support their contention for the existence of a common antigen between *R. conorii* and *R. prowazekii*. No *R. conorii* agglutination tests were reported. Giroud and colleagues also presented no data supporting the diagnosis of boutonneuse fever in the 11 persons who provided serum samples. However, on the basis of this work (and presumably more extensive unpublished data), they admitted that their *R. prowazekii* microagglutination test was not specific enough to distinguish between *R. prowazekii* and *R. conorii* antibody.

In spite of its admitted nonspecificity, they continued to use the microagglutination test with the crude *R. prowazekii* antigen and reported a high prevalence of *R. prowazekii* antibody in domestic animals in various African countries. In their first two papers, Giroud et al.[83,84] stated that the antibody in domestic animals was merely cross-reacting *R. conorii* antibody, even though they apparently did not test the positive sera with *R. conorii* antigen. Then to confuse the issue even more, Giroud and Jadin[85] did an about face and interpreted the results obtained on domestic animals from Ruanda-Urundi (now Ruanda and Burundi) as indicative of *R. prowazekii* infection. In this study they also did complement-fixation tests on some of the sera, but apparently discounted the almost complete lack of correlation with results of the microagglutination test.

The microagglutination test devised by Giroud and Giroud[86] has been widely used in Africa and to a lesser extent in South America and elsewhere but has never been critically evaluated. Few investigators using this test in serological surveys attempted to determine the specificity of the reactions they so dutifully recorded. Usually they did not use even the simplest controls or did not report them if they had. It is doubtful that the microagglutination test using crude antigens is able to distinguish between antibodies against any of the related tickborne and insectborne *Rickettsia* spp. The short paper of Zarate Aquino and Varela[88] purports to show that the microagglutination test can distinguish between fleaborne and louseborne typhus antibodies, but it does nothing of the kind. No other antigens were used, and absolutely no controls were mentioned. The sera tested were not described except that they came from various parts of Mexico and all had *Proteus* X 19–agglutinating antibodies. The widespread, uncritical use of such an unevaluated antigen and the liberal interpretations of the results have greatly confused the picture of possible louseborne typhus in domestic animals.

Fiset et al.[89] reported the development of a microagglutination test, which apparently does reliably measure rickettsial agglutinins in sera if it is used properly. The authors are careful to stress that the test is highly specific *only when highly purified antigens are used*. Crude or partially purified antigens are broadly cross-reactive.

b. Studies in Ethiopia and Egypt

Soon after the French accounts of typhus antibodies in domestic animals in Africa, Reiss-Gutfreund[67] reported the isolation of *R. prowazekii* from domestic animals and ticks in Ethiopia. This short paper was followed by a very detailed account of her work,[68] in which she tested several hundred serum samples from humans and domestic animals for agglutinating antibody to *R. prowazekii, R. typhi,* and *R. conorii* using the Giroud test. The results indicated that titers of $< 1{:}640$ with any of the three antigens were nonspecific. Those sera with titers $\geq 1{:}640$ were considered to be from animals that had been infected with the corresponding rickettsia. It was also discovered that 7% of 544 human sera and 17% of 649 cow, sheep, and goat sera had *R. prowazekii* agglutinating antibody titers $\geq 1{:}1280$. Considering the crudeness of the antigens and their great nonspecificity, these data would still be highly suspect were it not for the fact that *R. prowazekii* was actually reported to have been isolated from the blood of one goat and two sheep.

In a follow-up study, Reiss-Gutfreund[66] repeated her earlier results, demonstrating this time that 69% of 247 cattle, sheep, and goats from Ethiopia had *R. prowazekii*-agglutinating antibody at titers of 1:640 or greater. She apparently was also able to isolate additional strains of *R. prowazekii* from goats and from *Amblyomma variegatum*.

Unfortunately, attempts to reisolate the agents from the original material (a procedure that should be automatic for proper validation of any isolations) were not reported.

These startling results were not immediately confirmed by other workers, and the scientific community remained skeptical. Then Imam and Alfy,[90,91] reported complement-fixing antibody to *R. prowazekii* in the blood of numerous sheep, goats, camels, water buffalos, pigs, and donkeys in Egypt. The presence of specific *R. prowazekii* antibodies in some of these sera was confirmed by Lackman at the Rocky Mountain Laboratory.[92] Subsequently, three strains of *R. prowazekii* were isolated from donkeys,[93] and two of them produced significant rises in specific *R. prowazekii* antibody titer in guinea pigs using washed antigens.[94]

Reiss-Gutfreund and American workers[95] then collected more ticks and sera from domestic animals in Ethiopia for testing at the Rocky Mountain Laboratory in an attempt to repeat the earlier Ethiopian work. The results were completely negative for *R. prowazekii*, but *R. conorii* was found widely distributed in the Ethiopian tick population; and Burgdorfer et al.[96] obtained direct evidence of the presence of spotted fever group rickettsia by the hemolymph test in a high percentage of Ethiopian cattle ticks, but there was no evidence of typhus rickettsiae.

In 1966, Reiss-Gutfreund reported a further isolation of *R. prowazekii* from immature *Hyalomma* spp. ticks collected from cattle in Nazareth, Ethiopia.[42] She did not mention any attempt to reisolate the agent from the original tick suspension. However, this time the identity of the isolate was confirmed from yolk-sac material sent to the Rocky Mountain Laboratory.[97] Attempts by other investigators to isolate *R. prowazekii* from large numbers of Ethiopian ticks have been totally unsuccessful.[32,97–99]

In a later study in Ethiopia, Reiss-Gutfreund[100] failed to find significant antibody titers to *R. prowazekii* in domestic animals; also she was not able to isolate typhus rickettsiae from animal or human blood. She interprets these findings as indicating the possible seasonal or cyclic nature of *R. prowazekii* among domestic animals on the Ethiopian high plateau. At least it was less prevalent than previously reported. *Rickettsia conorii* appeared to have "taken its place" as the predominant rickettsial agent in domestic animals. One cannot help but wonder if *R. conorii* was not there all the time.

Subsequent work by a number of investigators has failed to clarify the relationship of *R. prowazekii* to ticks and nonhuman vertebrates. Philip et al.[95] and Burgdorfer et al.[96] attempted to repeat Reiss-Gutfreund's work, but failed to find any evidence of *R. prowazekii* infection in more than 6000 Ethiopian ticks. Russian workers failed to isolate *R. prowazekii* from ticks in several regions of the USSR, and were unable to demonstrate antibody in domestic animal sera.[101–103] Philip et al.[104] failed to find louseborne typhus antibody in any domestic animals tested in several South American countries although typhus antibodies were demonstrated in humans from most of the areas sampled. On the other hand, Varela and Velasco[105] reported the presence of agglutinating antibodies against *R. prowazekii* in domestic animals in Mexico.

In yet another study, Schmatz et al.[106] found typhus group CF antibody in four cattle and one goat from Jordan and two sheep from Egypt. Using type–specific agglutinating antigen, the antibody was identified unequivocally as *R. typhi*, not *R. prowazekii*.

5. Antibody in Wild Vertebrates

Hoogstraal et al.[107] reported the occurrence of group typhus agglutinating antibody in spiney mice (*Acomys* spp.) in Egypt, but the titers were too low (1:4 to 1:16) to permit specific identification of the antibodies as louseborne or fleaborne typhus. These results were confirmed the following year by Ormsbee et al.,[32] who found low levels of typhus group agglutinins in many sera from wild animals in Egypt. In three cases (two *Rattus* sp. and one *Arvicanthis* sp.) the diagnosis of specific epidemic typhus antibodies was unequivocal according to the authors. Attempts to isolate *R. prowazekii* from animal tissues and ticks were not successful. Varela et al.[108] reported complement-fixing antibodies against *R. prowazekii* in Mexican rats, but their findings are difficult to evaluate since they gave no data. It would be surprising if these were not *R. typhi* antibodies.

Marchette,[33] using a commercial complement-fixing antigen of the murine-epidemic typhus group, found low-titered antibody in the sera of only 2 of 471 domestic animals (cattle, goats, and pigs) in Malaysia. The two positive sera were from cattle and titered 1:4 and 1:8. Since epidemic typhus is unknown in that country, the antibody was not considered to be due to infection with *R. prowazekii*. In more than 1100 sera from wild mammals (mostly rodents, shrews, and tree shrews), typhus antibody was found only in rodents and shrews (*Rattus* and *Suncus*) associated with man. The presumptive isolation of *R. typhi* from *Rattus* and *Suncus* and the later definitive isolation (and reisolation) of a strain of *R. typhi* from *Rattus rattus diardii* from a dairy farm (unpublished) strongly suggest that the antibody was due to this agent and not to *R. prowazekii*.

Juminer[109] reported 20% of 400 *Rattus norvegicus* in Tunis with agglutinating antibody against *R. prowazekii*. In most instances little or no cross-reaction with *R. conorii, C. burnetii,* or *R. typhi* antigen was observed. These results differ considerably from those of other workers using Giroud's microagglutination test, but the reasons are not apparent. Rats collected in autumn and winter were much more likely to have *R. prowazekii* antibody than those collected in warmer months. A similar seasonal distribution of antibody in domestic animals has been noted in Egypt and Ethiopia.

6. Experimental Infection of Domestic Animals

Critical analysis of all information now available (1980) suggests that livestock in some regions, and perhaps wild animals as well, under certain circumstances, may become infected with *R. prowazekii* (and *R. typhi*). What the circumstances are and how infection is acquired have not been elucidated, and attempts to infect domestic animals experimentally have produced conflicting results. As early as 1938, Mosing[110] found that sheep in Poland possessed *R. prowazekii*-agglutinating antibody and were susceptible to experimental infection, but he did not consider his findings conclusive. Later, Reiss-Gutfreund[68] reported a febrile response in newborn lambs but not in adult sheep and cattle inoculated with large doses of *R. prowazekii*. Agglutinating antibody developed and persisted for some time in the inoculated animals. Immature ticks fed on experimentally infected lambs during the febrile period reportedly picked up the rickettsiae since *R. prowazekii* was isolated from the brain of a lamb several weeks after inoculation with the tick suspension. In subsequent experiments, she reported the isolation of *R. prowazekii* from the blood of experimentally infected lambs using rectal inoculation of human lice, but not by inoculation of guinea pigs or embryonated eggs.[111] In similar experiments, Sarycheva and Chirov[112] were unable to infect lambs.

Philip et al.[113] were unable to detect a rickettsemia or clinical response in experimentally infected goats, calves, and a donkey; but they did confirm the development of specific *R. prowazekii* antibody when large inocula were given. These results were confirmed in extensive experiments by Ormsbee et al.,[92,114] who also failed to produce infections in young sheep, goats, camels, and donkeys, even when massive doses of *R. prowazekii* were given by several routes. Their animals did develop antibodies, but the antibody response varied with the size of the inoculum and could be elicited equally well with killed organisms. They were unable to isolate *R. prowazekii* from peripheral blood or other tissues from any of the inoculated animals and concluded that antibody response alone is not certain evidence of infection in domestic animals. Massive intraperitoneal and intravenous inoculation of *R. prowazekii* into lambs produced limited short-term infections of internal organs, but no rickettsemia or lymph node infection.[92]

A short paper by Giroud[115] claimed to demonstrate the development of antibody (presumably agglutinating) in a horse inoculated with a strain of *R. prowazekii* reportedly isolated by Reiss-Gutfreund from a horse in Ethiopia. The organism was avirulent for guinea pigs, and no data were given establishing the organism as *R. prowazekii*.

The evidence from experimental infections fails to support the hypothesis that domestic animals are a significant part of any extra-human *R. prowazekii* cycle. The explanation for serological evidence of natural typhus infection in livestock in Africa is not readily apparent.

7. Experimental Infection of Ticks

The evidence implicating ticks in a natural, nonhuman cycle of *R. prowazekii* is still entirely circumstantial. The reported isolations of *R. prowazekii* from ticks in Ethiopia does not necessarily imply that the organism can persist in them and eventually be transmitted to other vertebrates. Reiss-Gutfreund's earliest report[67] indicated that immature ticks of several species could be infected by ingesting blood of infected rabbits and lambs, but the adults could not. Grokhovskaya et al.[116,117] demonstrated that *Hyalomma anatolicum, Dermacentor pictus,* and *Alectorobius* (=*Ornithodoros*) *canestrinii* picked up the rickettsia during feeding on rickettsemic guinea pigs and remained infected for 15 days. Attempts to transmit the infection by feeding infected ticks on healthy guinea pigs were not successful. Other Russian workers have tested a number of argasid and ixodid ticks for susceptibility to infection with *R. prowazekii* acquired by ingesting infective blood. In argasid ticks, *R. prowazekii* reportedly persisted for at least 9 months and was passed transstadially. Klimentova and Perfil'yev[118] and Ignatovich and Grokhovskaya[119] reported transmission of *R. prowazekii* to guinea pigs, but the rate of success was very low. The rickettsia persisted for shorter periods in ixodids, and species of *Dermacentor* were more readily infected than species of *Hyalomma*.[120] Burgdorfer and Ormsbee,[98] on the other hand, were unable to infect *Hyalomma dromedarii, H. anatolicum excavatum,* or *Dermacentor andersoni* either through glass capillaries or by feeding them on rickettsemic meadow mice or gerbils. Later, Burgdorfer et al.[121] demonstrated short-term infection of these and other ticks, but the organisms were restricted to gut tissue and the ticks were unable to maintain or transmit the agent. Of purely historical interest is the early report of Zinsser and Castaneda[122] that *R. prowazekii* injected rectally in *D. andersoni, D. nitens,* and *Amblyomma* sp. remained alive for at least 14 days. Weyer[123] was able to maintain strains of *R. prowazekii* for several years in *Ornithodoros moubata,* infected either intracoelomically or by feeding on rickettsemic guinea pigs. The rickettsiae multiplied in all tissues and were passed through eggs to the offspring. To my knowledge, these results have not been confirmed in other laboratories, but Grokhovskaya et al.[120] reported that *O. moubata* is susceptible to the rickettsia in all stages and demonstrated transstadial transmission. They and Ignatovich and Grokhovskaya[119] also demonstrated that *O. moubata* could experimentally transmit *R. prowazekii* to animals by bite but other ticks could not. In an earlier report Grokhovskaya et al.[116] failed to transmit *R. prowazekii* by any of a number of ixodid and argasid ticks including *O. moubata*. These authors suggest the possibility of a significant relationship between *R. prowazekii* and argasid rather than ixodid ticks but their data are not convincing.

In later experiments, Ignatovich and Grokhovskaya[124] demonstrated that various species of ticks can be infected by feeding on rickettsemic guinea pigs or through a membrane, but that viable rickettsia persisted in them for only short intervals unless they ingested large numbers of organisms. They also showed that infection could be transmitted to susceptible animals by rubbing crushed infected ticks into scarified skin and, rarely, by bite. These workers did not consider their results to be definitive proof that ticks are involved in natural typhus cycles. Ormsbee et al.[92] were even less successful in infecting ticks (both ixodids and argasids) by feeding them on infected voles and gerbils which produce longer and higher titered rickettsiemias than the rabbits or guinea pigs used by other investigators; nor were they able to demonstrate transmission by tick bite. They concluded that the data available suggest that the minimal dosage of rickettsia necessary to infect ticks must be extremely high. The inability to detect rickettsemia in domestic animals[92,113] seriously questions the hypothesis that domestic animals and ticks are involved in natural *R. prowazekii* cycles.

Similar conclusions were drawn by Burgdorfer et al.[121] in yet another series of experiments. They demonstrated limited persistance of *R. prowazekii* in ticks after feeding on rickettsemic voles, but in all cases the organism was confined to the tick gut. No generalized infections, as regularly seen in spotted fever infections, were ever seen except in ticks inoculated intracoelomically. Tick hemolymph seems to be a suitable medium for the multiplication of many rickettsiae by inoculation[98,123,125,126] although louse hemolymph is probably better. One strange phenomenon that has not been explained is the original observation by Piontkovskaya and Korshunova[127] that most ticks of the genus *Dermacentor* die after acquiring infection with *R. prowazekii*. This was confirmed for *D. andersoni* by Burgdorfer and Ormsbee[98] and Burgdorfer et al.[121] The fact that *R. prowazekii* and *R. typhi* multiply in *Hyalomma dromedarii* tissue cultures[128] is of unknown significance, but this may be a simple way to study at least one aspect of tick-rickettsia interaction. *R. conorii* and *R. akari* grew better in the tick cells than did either of the typhus rickettsiae.

8. Hypotheses Concerning Origin of Typhus Antibodies in Domestic Animals

The hypothesis that ticks are involved in a *R. prowazekii* cycle involving domestic or wild animals has not been supported by extensive experimental work. The insensitivity of ticks to infection with all but very high concentrations of *R. prowazekii* seems to eliminate domestic animals as the source of infection of those ticks reported to be naturally infected in Ethiopia. One cannot completely rule out the possibility that some wild–animal host may produce rickettsial concentrations in their blood high enough to infect ticks, but there is no evidence whatsoever for the existence of such an animal.[121]

An alternative hypothesis by Giroud et al.[129] to explain the presence of *R. prowazekii* antibody in domestic animals suggests that it and *R. conorii* possess a common antigenic component, and that under certain conditions one may be transformed into the other. The evidence for this, however, is fragmentary, and the ability of *R. prowazekii* to change into *R. conorii* or vice versa has certainly not been demonstrated. Along the same line, Reiss-Gutfreund,[130] noting that *R. prowazekii* was gradually replaced by *R. typhi* in Ethiopia, suggested the possibility that *R. prowazekii* had mutated to *R. typhi*. Evidence for this is lacking, although eons ago the reverse mutation or series of mutations probably occurred.

It is well known that *R. prowazekii* shares antigens with *Proteus* bacteria, and it is conceivable that Egyptian and Ethiopian livestock may be infected with microorganisms that share antigens with *R. prowazekii*. This has been suggested by Sarycheva[131] and by Pecic and Borjanovic[132] who found *R. prowazekii* CF antibodies to be as prevalent in Serbian sheep in regions where typhus had not existed for years, as in sheep in hyperendemic regions. Until such organisms are isolated and identified, however, the existence of typhus antibodies in livestock remains to be explained.

In some endemic or hyperendemic areas, domestic animals, particularly those in close contact with man, could possibly become infected with *R. prowazekii*, but the available evidence does not support the hypothesis that ticks are involved in transmitting the agent to or among domestic animals. Reports of isolation of *R. prowazekii* from ticks have come only from Ethiopia and have not been confirmed. Philip and Imam[93] suggested that domestic animals are originally contaminated with typhus by man, but the mechanism of infection is still unknown. Reiss-Gutfreund[130] has noted the positive correlation between prevalence of *R. prowazekii* antibody in man and domestic animals in Ethiopia and suggests that livestock acquire their infection from man via human lice or louse feces. It is conceivable that domestic animals might become infected secondarily with either *R. typhi* or *R. prowazekii* by inhalation or by ingestion of infective flea or louse feces. The regions in which the serologically positive animals are found are all hyperendemic typhus areas. It is also the local custom of the peoples there to live in close association with their livestock. Repeated exposure of cattle and sheep to highly infectious louse feces might result in the production of typhus antibody. Experimental pulmonary infection of *R. prowazekii* in dogs, sheep, and goats has been reported by early investigators,[133–137] although Ormsbee et al.[92] failed to produce infection by intratracheal inoculation.

The confusing story of epidemic typhus ecology in Ethiopia and Egypt still needs clarification. All indications are that the ecological situation there is peculiar to that part of the world for reasons that may be related to the high endemicity of typhus in the human populations. Unless or until more convincing evidence is presented, it must be concluded that the apparent infection of domestic animals and certain of their ticks is purely accidental, is of no phylogenetic significance, and is of only limited ecological importance.[99,121,138–140]

9. Maintenance of *R. prowazekii* in Man

The postulation of a nonhuman interepidemic maintenance system for *R. prowazekii* is probably not necessary to explain its survival between epidemics. The tendency for the organism to remain latent in patients long after recovery from the acute disease is well known. Reactivation of such latent infections may result in a second attack of typhus, the so-called recrudescent typhus or Brill-Zinsser disease. Weyer[141] is of the opinion that man is the only vertebrate-maintaining host of the typhus organism. The work of Murray et al.[142] and Gaon and Murray[143] on Brill-Zinsser disease in the typhus zones of Yugoslavia strongly supports Zinsser's hypothesis that man is the interepidemic maintaining host of epidemic louseborne typhus fever. Typhus epidemics in endemic areas often begin with the appearance of atypical cases whose symptoms are much milder than succeeding ones.[110] The source of these first cases is usually not known. In contrast, epidemics in nonendemic areas are usually caused by introduction of the disease by travelers. Mosing's work[110] also suggests that lice do not always die within a week or two after infection with *R. prowazekii*, contrary to almost all textbooks. If lightly infected and if the temperature is a few degrees below the optimum (37° C), infected lice may live 2 or 3 months. These observations suggest that cases of recrudescent typhus may be responsible for initiation of epidemics and that man may be considered as the interepidemic maintaining host of *R. prowazekii*.

Studies in Eastern Europe suggest that recrudescent typhus is still prevalent.[144–149] Equally important are the observations that lice can transmit typhus from individuals with Brill-Zinsser disease to susceptible individuals,[145,147] even though the infection rate among lice fed on these patients is generally much lower than in those fed on classical typhus cases.[150–152]

III. ORIGIN OF TYPHUS RICKETTSIAE

A. Early Ideas on the Relationship of *R. prowazekii* to *R. typhi*

The isolation of a strain of *Proteus vulgaris* from the urine of a typhus patient, and the finding that it was agglutinated by the patient's serum, and that of other typhus patients,[153] provided the first clues to the possible origin of *R. prowazekii*. These and further observations led Weil and Felix[154] to propose that the parasitic typhus rickettsiae are exceptionally, and under unknown conditions, transformed into the culturable stage known as *Proteus* X. The occurrence in Malaya and Sao Paulo of other *Proteus* X strains closely related to the disease produced by the particular typhus rickettsia in these widely separated places lent support to the hypothesis.[155] But many difficulties could not easily be explained away. Not all typhus patients produced good *Proteus* X agglutinins, and experimentally infected guinea pigs and other animals also often failed to do so. A growing awareness was eventually expressed by Castaneda[156] that the mere presence of a common antigen between organisms does not necessarily indicate relationship. He demonstrated that while the sera of rabbits immunized against *Proteus* X 19 fixed complement with *R. prowazekii* and with *Proteus* antigens, typhus-immune rabbit serum contained CF antibody only to *R. prowazekii*.

Another outstanding feature of typhus that struck the investigators of the early 20th century was the apparent lack of a primary animal host. Morphologically the typhus organism was nearly identical to the tickborne spotted fever organism, but its only known arthropod host was an obligate human parasite, the human body louse. This was all to change with Maxcy's findings[157,158] that an endemic form of typhus in the U.S. was not naturally transmitted from man to man by the louse, although Mooser and Drummer[159] showed that the louse could transmit it experimentally. He suggested that, except for its mildness and epidemiology, this disease was indistinguishable from classical typhus, that it had a common origin with the disease called *tabardillo* in Mexico,[160] that the probable maintaining hosts were rats and mice, and that man acquired the disease accidentally by bite of some bloodsucking parasite. Almost simultaneously, typhus rickettsiae were isolated from fleas in Baltimore[161] and from the brains of rats trapped in Mexico City.[162] Similar isolations were soon made from rats and fleas in many parts of the world, and Mooser[163] named the disease murine typhus. Even earlier, he suggested that the Mexican *tabardillo* is distinct from European typhus and identical with the disease in the southern U.S.[164]

The murine or fleaborne typhus rickettsia, although similar to the human or louseborne form, could be distinguished from it, and it was soon recognized that the two organisms are closely related but distinct entities. It was then only a matter of time before alert minds began to realize that one form was probably derived from the other. Mooser[169] was of the opinion that the louse was probably not the original vector of *tabardillo* in Mexico because it is lethally susceptible to the infection. Later, he and Castaneda[165] demonstrated that *R. typhi* replicates abundantly in the intestinal epithelium and Malpighian tubules of several fleas (*Xenopsylla cheopis, Ceratophyllus fasciatus, Leptopsylla musculi, Ctenocephalus canus*, and *C. felis*), but does not grow in the mites *Laelaps echidninus* or *Liponyssus bacoti*. He postulated that typhus was originally and exclusively a disease of rodents from which it has been transmitted to man by fleas or possibly by some variety of tick.

Hypotheses once formed, however, die hard, and we find two of the foremost typhus workers in 1935 still suggesting that *B. proteus* is the ancestor of the rickettsiae then known[166] or, conversely, that the insectborne rickettsiae are the ancestors of *B. proteus* X 19.[167]

In 1943, Baker[168] classified Mexican typhus (*tabardillo*) as a form intermediate between classical louseborne typhus and fleaborne murine typhus. According to his typical epidemic series, it is a Class 4 disease in which a secondary arthropod, the louse, becomes involved in transmission. Thus murine typhus agents infecting man are ingested by the louse, which in turn continues the transmission cycle from louse to man to louse; the virulence of the rickettsia thereby increases, and an epidemic situation arises. Louseborne "murine" typhus epidemics, however, are sporadic and self-limiting, because the rickettsiae never have been able to develop a completely independent existence in man and louse. A similar situation may exist in South America, although Zinsser[169] thought that *R. prowazekii* might also occasionally occur there. We now know that both classical louseborne typhus (caused by *R. prowazekii*) and sporadic, fleaborne typhus (caused by *R. typhi*) exist in Mexico.

The disease called *tabardillo* is thought by some to be a louseborne, epidemic form of fleaborne typhus,[169–171] a possibility that deserves more attention than it has received in recent years. According to the authors, the disease that the Spanish called *tabardillo* was already known to the early Aztecs of Mexico as *Matlalzahuatl* before the Spanish arrived.[172] It was supposed to be transmitted by a trombiculid mite (*Trombicula irritans*) that they had named *Tlalzhuatl*. Although Sambon considered *Matlalzahuatl* to have been classical epidemic typhus, he also recorded that the Spanish referred to the disease as *sarna en el rebono*—the itch or eruption of the flock or crowd. In all probability, *Matlalzahuatl* was originally the Aztec name for the syndrome produced by heavy infestations of the itch mite, *T. irritans*. It did not refer to *tabardillo* or typhus until after the arrival of the Spaniards when it was apparently also used for typhus fever. The worst typhus outbreak of all time in Mexico City lasted from 1576 to 1581.[173] Few Spaniards were affected, but the Indians died by the thousands, suggesting that the Aztec population had had no previous experience with this disease.

It is also possible that the Aztecs did not distinguish between the rash produced by *T. irritans* and that produced by fleaborne typhus. It is highly unlikely that the Spaniards introduced *R. typhi* as well as *R. prowazekii* into Mexico. If fleaborne typhus was already endemic among the Aztecs when the Spaniards arrived, the apparent absence of epidemics of typhus suggest that the conditions of early Aztec life were not conducive to rapid transmission of *R. typhi* in epidemic form, or that the human louse was not there. Introduction of the louse by the Spaniards provided the means by which fleaborne typhus could become epidemic through rapid man-to-man transmission. That the Spanish *conquistadores* brought the human louse with them is certain, but there is some evidence that it was already there (see III,C,6, this chapter). Castaneda has suggested that most great past epidemics in Mexico since the Spanish conquest probably were classical louseborne typhus caused by *R. prowazekii,* which is now endemic in most temperate and cold regions of the country. The most reasonable hypothesis is that *tabardillo* was a term used by the Spaniards to refer to epidemic louseborne typhus which they had unwittingly introduced to the Aztecs. In more modern times, louseborne *R. typhi* may also have been responsible for epidemics of high mortality in Mexico City, where both forms of typhus occur, because of such factors as relaxation of sanitary measures, crowding, and undernourishment during times of social stress. However, from 1939 to 1948, practically all strains isolated were *R. prowazekii,* suggesting that epidemics of louse-transmitted *R. typhi* may not be very common.

B. Evolution of *Rickettsia typhi*

1. Early Thoughts on Place of Origin

Nicolle,[174] one of the first to speculate on the subject, placed the origin of fleaborne typhus in Asia whence it was carried to North America in rats on Polynesian vessels. From there it was carried by boat to European ports in historical times. Heyerdahl[175] dramatically demonstrated that the voyage in the opposite direction, from Peru to Polynesia, could be made in a simple raft. Fifty years earlier, Captain Voss[176] sailed a 38-ft Northwest Indian dugout canoe (named "Tilikum") from Vancouver to Tongareva in Central Polynesia in only 2 months with a watch and a compass as his only navigational instruments. Heyerdahl[177] presented evidence for the initial settlement of Polynesia by peoples from Peru, but his hypothesis has not gained wide acceptance among anthropologists, most of whom believe that Polynesia was settled by people from Southeast Asia migrating in a southeasterly direction along the island chains. Much later, however, there probably was cultural exchange between Polynesia and South America by movement of peoples in both directions.*

By whatever route *R. typhi* reached the New World, it probably did so in historic times as a parasite of *R. rattus* and its ectoparasites carried aboard ships. This readily explains the present world-wide distribution of *R. rattus* and *R. typhi*. The accepted origin of *Rattus* in Southeast Asia strongly suggests that *R. typhi* also originated there. It is unlikely that *R. rattus* came to the New World via the South Pacific for reasons discussed in a later section. Instead, it probably was brought to North America by the early explorers in the 16th century,[182] and introduced into South and Central America about the same time;[183] at least it was not recorded in South America before 1544.

Baker[168] speculated that the original home of typhus was the eastern end of the Mediterranean, based on the presumed origin of *Xenopsylla cheopis* in the Nile Valley. But it should be remembered, as mentioned previously, that Baker was primarily interested in the origin of the disease in man, not in the parasite itself. These are two quite different things. He was well aware of this and even considered it likely that *R. typhi* originated as an acarine parasite and that *X. cheopis* was not the primary arthropod but that ". . . it has merely usurped the role of such an arthropod in animal and human transmission." If the mite, *Ornithonyssus bacoti,* is the primary arthropod host (and Baker does not accept the evidence without reservation), the hypothesis would not be materially changed because, ". . . the mite was first recorded from Egypt and has migrated to many parts of the world, including the United States. It too could place the cradle of typhus near the eastern end of the Mediterranean." The fallacy of this line of reasoning is that the basic assumption, the origin of *O. bacoti* in Egypt, is probably incorrect. The place at which a species is first recorded does not necessarily have any relation to its center of origin. In this particular instance, the first collection of *O. bacoti* actually was made in the U.S., not Egypt, although it was first *described* from material collected in Egypt. But more important, the association of *O. bacoti* with *Litomosoides carinii,* a filarial parasite of the cotton rat *Sigmodon hispidus,* provides good evidence of its New World origin. This nematode apparently is transmitted only by *O. bacoti* and is primarily a parasite of *Sigmodon,* a New World rodent.[184]

2. *Mites as Original Hosts*

Huff's[185] early hypothesis that *R. typhi* originated as a symbiote of free-living mites was based primarily on the reports by Dove and Shelmire[51,52] that the rickettsia may be transovarially transmitted in the rat mite *Ornithonyssus bacoti*. In their experiments,

*The field of ethnobotany sheds some light on the origin of the Polynesian peoples and strongly indicates that they were Indo-Malaysian emigrants. The Indo-Malaysian region is the center of origin of various edible plants now cultivated throughout the area.[178] As one proceeds in an easterly direction into Oceania, the number of cultivated species decreases, suggesting that the Proto-Polynesian emigrants followed a route from Indonesia through either Micronesia or Melanesia.[179] Many other lines of evidence point to the Indo-Malaysian (and ultimately Asian) origin of the Polynesians, who subsequently had cultural contacts with Central and South American Indians.[180,181]

infected mites were placed in tubes and allowed to lay eggs, which later hatched into larvae. The female mites were removed and the larvae were placed on a guinea pig to feed. The guinea pig subsequently developed typhus infection. Unfortunately, no attempt was made to free the larvae from external contamination with the infectious feces that the female mites undoubtedly excreted into the tube. Mooser et al.[186] were unable to confirm these results and could never demonstrate any multiplication in mites (or ticks), nor could they accomplish transmission with either. The experiments of Worth and Rickard[187] suggest that this mite is, at best, an inefficient vector even in the infected adult stage. In more recent work, Zemskaya[188] reported that *R. typhi* is transmitted transovarially in *O. bacoti,* but to my knowledge this has not been substantiated.

The question of transovarial transmission is of considerable biological and evolutionary importance and one that could be solved relatively easily with the newer techniques using specific fluorescein-tagged antibody as described by Burgdorfer.[189] Unless and until it can be conclusively shown that *R. typhi* is naturally transmissible through the ova of rat mites, the role of these ectoparasites as basic hosts of the rickettsiae in natural cycles of infection remains doubtful.

Whatever the present role of the rat mite in the ecology of murine typhus may be, it has been secondarily acquired. All but 2 of the 15 or more species of *Ornithonyssus* are parasites of rodents and other small mammals. The other two are bird parasites. *Ornithonyssus bacoti* and *O. ondatrae* have been widely dispersed by human activity and are found in many parts of the world. All the others are known only from the New World,[190] and there is evidence that the genus evolved in the Western Hemisphere.[184] *Ornithonyssus bacoti* is cosmopolitan in distribution within the tropical and temperate zones.[184] It is thought to have evolved in the neotropics and to have been distributed throughout the world by domiciliated rats after they were introduced into the New World by man in recent times and carried with him in his travels.

3. Xenopsylla cheopis

Much significance is often assigned to the lack of hereditary transmission of *R. typhi* in *Xenopsylla cheopis,* and this fact is used to support the thesis for its recent divergence from a tickborne progenitor. However, the inability of *R. typhi* to cross the gut wall barrier* in the flea does not necessarily indicate a recent origin and poor adaptation to its arthropod vector. Transovarial transmission of microbial symbiotes is highly developed in many insects as well as in the Acari, but there is no solid evidence that other

*The term "gut wall barrier" is another instance of the inappropriateness of borrowed terms. The implication is that there is a specific structure, formation, or configuration—a barrier—in the gut wall, which, by its presence, prevents the organism from passing through into the peritoneum or hemocele. The so-called barrier does not exist in the gut anymore than it does in the parasite itself. Such terms are useful to the specialist when their meaning is well understood. In general, however, they give the impression, particularly to the student and nonspecialist, that more is known about the subject than is actually the case. Gut-wall barrier does not refer to any definable property of gut walls in general or any gut wall in particular, which prevents organisms from crossing them. What might be a barrier for one organism is an open passageway for others. Usually we simply do not know why some organisms inhabit the lumen of the gut, while others reside in the epithelium, and still others infect the gut cells only temporarily on their way through to the peritoneal cavity (or hemocele) or to some other organ system via the lymphatics, blood vessels, neurones, or tissue spaces. The gut lumen is an ecosystem as are the gut epithelium, the entire digestive system, the peritoneum, the hemocele, the central nervous system, the whole animal body. Each parasite, symbiote, or commensal inhabits the particular ecosystem to which it has become adapted, and the activities and population growth of each is influenced and controlled by interaction with the other elements of its ecosystem and those of surrounding ecosystems, which have evolved over long periods of time. The term barrier implies that the process is a static one rather than the dynamic equilibrium that it actually is.

symbiotes, such as arboviruses in mosquitoes, plague bacilli in fleas, leishmania in sandflies, filariae in black flies and tabanids, spirochaetes in lice, etc. are naturally transmitted hereditarily in their insect hosts. This does not mean that they are all of recent origin. Some plant viruses seem to readily pass through the gut wall and infect the oocytes of some insects, but the evidence that many vertebrate viruses do so is not entirely convincing. Transovarial transmission of Japanese encephalitis virus,[191] La-Crosse virus,[192,193] and yellow fever virus[194] have been demonstrated in the laboratory, but it is not known if this occurs widely in nature. In the phlebotomous fever virus group, transovarial transmission in the vector flies appears to be the rule,[195] and it may also be a natural mechanism of survival in the California group of viruses. Adult mosquitoes reared from field-collected larvae have been found to be infected with California encephalitis virus,[196] and a number of other serotypes of this virus group have been isolated from developmental stages of vector mosquitoes collected in the field.[196]

It is not likely that *R. typhi* originated as an insect parasite but its relationship with rat fleas and lice may be older than is usually believed. *R. typhi* is more closely related to the tick typhus (*Dermacentroxenus*) than to the trombiculid mite (*Zinssera*) stem of the genus *Rickettsia,* indicating that it branched from the main *Dermacentroxenus* trunk long after the divergence of *R. tsutsugamushi*. The recently discovered *R. canada* is the tickborne agent to which *R. typhi* is most closely related. Little is known about *R. canada,* but this close antigenic relationship does support the hypothesis that *R. typhi* originated from a tickborne ancestor.

4. *The Rat*

The adaptation of *R. typhi* to rats is so complete that it produces only an inapparent infection, and the organism is capable of persisting in the rat's tissues for up to a year or more. The extremely long latent infection produced by *R. typhi* in rats and other rodents has been experimentally demonstrated many times.[197–200] Latent infection can also be produced experimentally by *R. prowazekii* in cotton rats,[201] but they are not naturally infected with this agent. The persistence of the agent in naturally infected wild rats is not known. However, *R. typhi* is more readily isolated from the brains of wild rats than from any other organ; this might suggest a latent infection since no signs of central nervous system disease are observed. Some neurotropic viruses also are known to have unusually long incubation periods, during which they may be in a "dormant" phase in the nervous tissue (for example, rabies, scrapie, visna). Other viruses retreat to nerve ganglia after the acute infection and may remain dormant for years before some stimulus sends them back down the axons and into susceptible tissues (for example, herpes simplex virus and herpes varicella-zoster virus).

The most logical hypothesis for the origin of *R. typhi* is that it was a tickborne rickettsia of *Dermacentroxenus* parentage, which became especially well adapted to its vertebrate hosts, species of *Rattus,* in Southeast Asia. The adaptation may have taken place most rapidly, or perhaps only, in *Rattus rattus* as it diverged from its forest relatives and became restricted more or less to the forest fringe and fields or scrub habitat.

Southeast Asia is the only region with an indigenous fauna of *Rattus rattus*. Many subspecies have developed where subpopulations have become isolated, as on the many islands of the South China Sea and Indonesian Ocean. *Rattus rattus* is usually restricted to open habitats and is not found in primary forests except where the usual jungle species are absent as on Pulau Jarak.[202] When primitive humans began clearing the forests to form settlements, the habitat of *Rattus rattus* was enormously expanded, and the rat, being compatible with its more recently evolved cousin, thrived as never before. For one reason or another, forest ticks are not well adapted to field and scrub habitat, and the rats were relatively tick free.[203] To survive in this ecosystem, a normally tick-

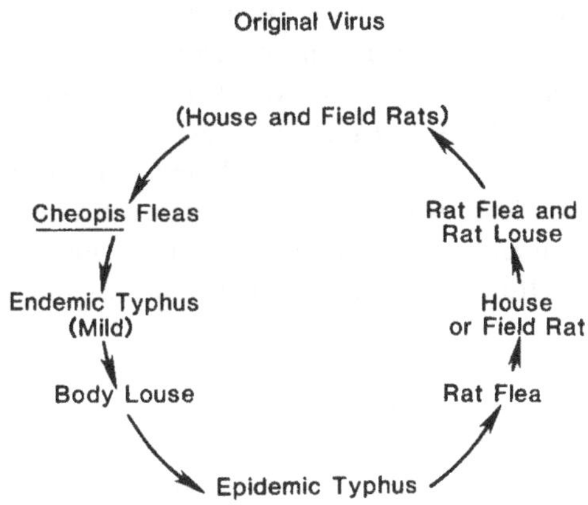

FIGURE 1. The *Rickettsia typhi-R. prowazekii* cycle according to Kodama et al.[204]

borne rickettsia had to either adapt more completely to its vertebrate host or find a new arthropod host. *Rickettsia typhi* apparently did both, but established a closer relationship to the rat than to its extoparasites. Both are probably necessary for the continuation of the cycle, but the rat is clearly the maintaining host. This rather unique adaptation to a vertebrate host rather than to an arthropod advanced one step further in the evolution of *R. prowazekii*.

C. Evolution of *Rickettsia prowazekii*

1. Derivation from *R. typhi*—Early Ideas

Three Japanese workers, Kodama, Kono and Takahashi[204] apparently were first to propose that *R. prowazekii* was derived from *R. typhi*. They isolated a rickettsia (which they named *R. manchuriae*) from *Rattus norvegicus, X. cheopis,* and *P. spinulosus* in Manchuria and suggested that *R. prowazekii* originates by passage of murine typhus through the human louse. They considered this a dynamic process in which the human form readily changes back to the murine form by passage again through the rat flea. Their proposed cycle is shown in Figure 1. A year later Lepine[205] expressed the opinion that epidemic typhus certainly arose from murine typhus by mutation, and the two still remain very closely related.

Investigators then attempted to demonstrate experimentally the conversion of one form of typhus rickettsia into the other,[82,186,206–209] all without success. Mooser et al.[186] thought they had converted a Mexican strain of epidemic typhus into a murine strain, but they were probably working with strains of murine typhus (initially nonorchitic in guinea pigs) all the time. Price et al.[210] reported the most elaborate and sophisticated attempts of all. Despite great technical difficulties, they showed that extracts of *R. typhi* could, under experimental conditions, transform *R. prowazekii* to *R. typhi*. The extract was principally deoxyribonucleic acid and the transformation was permanent. To my knowledge these experiments have not been repeated, and there is a reasonable doubt that Price et al. were able to completely eliminate the presence of viable *R. typhi* from their extracts.

In the absence of more convincing evidence than obtainable to date, we must conclude that *R. typhi* and *R. prowazekii* are distinct entities leading separate and now unrelated existences, each occupying its own niche. The discovery and characterization of the basic antigenic differences between the two rickettsiae[211–214] make it possible to readily and accurately distinguish between them. They also provide a qualitative and quantitative means for estimating the degree of relationship existing between the two species. Antigenic characters alone do not necessarily determine a species; but, when added to knowledge of its ecology and that of all other related forms, a reasonably reliable judgment of specific status and relationships can be made. This is true for *R. typhi* and *R. prowazekii*. They are sufficiently different to be accorded specific status, but their relationship is so close that a common origin can reasonably be postulated. *Rickettsia typhi* and *R. prowazekii* probably did not arise from a common ancestor, but one actually arose directly from the other. Furthermore this may well have happened in historical time.

As already discussed, *R. typhi* probably arose from an originally tickborne parent which became adapted to species of *Rattus*, probably in southern Asia, which in turn had become adapted to a field or scrub type of habitat. Man's sole contribution to the evolution and development of *R. typhi* was in providing its vertebrate hosts with a suitable environment for their expansion and eventual dispersion throughout the world. House shrews and mice may occasionally become infected with *R. typhi* but are not essential to its basic ecology. Species of *Rattus*, particularly *R. rattus*, are the primary hosts of murine typhus rickettsiae, and it is to these ubiquitous creatures we now turn to develop an hypothesis of the origin of epidemic typhus.

2. Origin and Dispersion of *Rattus rattus*

The original home and center of dispersal of *Rattus* is Southeast Asia (see Volume I, Chapter 4). This large genus is essentially restricted to Asia except for two forms, *Rattus rattus* subspp. and *R. norvegicus* (=*R. r. norvegicus*), which are worldwide thanks to their close ecological relationship with man, and *R. exulans* (=*R. r. exulans*) which has colonized the Pacific islands. Schwarz and Schwarz[215] consider *R. exulans* and *R. norvegicus* to be subspecies of *R. rattus* while admitting that the taxonomy of commensal rats is difficult to disentangle without determining the extent with which the different subspecies are able to reproduce hybrid offspring. Pincus[216] has shown that *norvegicus* will not produce offspring with *R. rattus*. This would seem to justify its specific status, and it will be referred to here as *R. norvegicus*.

The varieties of *R. rattus* (commensal black rats) now populating the world are derivable directly from the ancestral form, *R. r. diardii* Jentink in Southeast Asia,[217] which in turn developed from the wild Javan rat, *R. r. roquei*. According to Schwarz and Schwarz,[215] a population of wild rats in Java became segregated ecologically into three major subgroups: (1) *roquei* stayed in the bush, (2) *diardi* became adapted to human habitations, and (3) *argentiventer* went into plantations (rice, sugar cane, etc.) with access to water. The dark-bellied *diardi* rats have segregated into *R. r. tanezumi* in southern China; *R. r. alexandrinus* in northern and central India and the Mediteranean region; *R. r. rufescens* of southern India, Sri Lanka (Ceylon), Zanzibar, and Madagascar; *R. r. rattus* originally in northwestern Europe and now widespread throughout the world; and *R. r. diardii* in Malaysia and Indonesia. All these rats are found only in close association with man.

The silver-bellied rats of the rice fields and sewers all have a predilection for water and are primarily outdoor rats. They include *R. r. argentiventer* in Malaysia and Indonesia, *R. r. nitidus* in South China, *R. r. caraco* in East Asia, *R. r. kandianus* in Sri Lanka, and *R. norvegicus* in Europe and port cities around the world.

The *exulans* series of Malaysian and Pacific commensal rats developed in similar fashion from *R. r. wichmanni*, a close relative of *R. r. roquei* in the Lesser Sunda Islands east of Java. The most widespread is *R. r. exulans*, which has populated the South Pacific islands as far south as New Zealand and as far east as Hawaii. The other forms are restricted to Southeast Asia.

***a.* Migrations of *Rattus* to Europe and Beyond**

It is not known when *R. r. diardii* first migrated from its ancestral home, but it must have been through the agency of man with whom this rat had long been closely associated. Some of the early Indian traders, who brought Hinduism to the Malay Archipelago, may have been the first to ferry *R. r. diardii* back across the Indian Ocean to the southeast coast of India. From there early Arab sailors could have carried it to Africa and the Middle East.

However, there is some evidence from a few fossil remains that *R. rattus* may have migrated to Europe even earlier, but probably not before it developed a strong association with man. Pictet[218] described remains, apparently of *R. rattus* from Pleistocene gravels at Mattegnin, Geneva, but their actual age is questionable.[183] A few remains (mostly lower jaws) that have been referred to as *Rattus* were described from other European Pleistocene deposits.[219–221] Bate,[222] in discussing the fossil history of the genus *Rattus,* does not mention the European material although she does mention doubtful records from Gibraltar and Malta. An upper jaw from Pleistocene deposits on Crete differed significantly from that of either *R. rattus* or *R. norvegicus* and was described as a new species, *R. kiridus.*[222]

A single mandibular ramus from Mesolithic deposits in Israel was described as *Rattus (Mastomys)* sp. by Bate.[223] *Mastomys* is now considered a separate genus only remotely related to the *R. rattus* group. Ellerman[224] considered a large series of African Muridae as subgenera of *Rattus*. *Stochomys* is nearest to the typical *Rattus* of Asia; the remaining subgenera (*Praomys, Hybomys, Myomys, Mastomys, Dephomys,* and *Ochromys*) are smaller and more generalized than the Indo-Malaysian forms. The native African rats formerly ascribed to various subgenera of *Rattus* are now placed in distinct genera.[225]

Two mandibles from upper Pleistocene deposits at Kebara, Israel, have been assigned to *Rattus rattus* although differences from modern *R. rattus* were noted.[226] Natufian-Neolithic deposits from the Abu Usba Cave, Israel, have yielded two mandibles and two maxillae which are most similar to modern *R. rattus.*[227]

Haas[228] found several mandibles in the Oumm-Qatafa cave in Israel, which he placed in the genus *Rattus*. Tchernov[226] erected a new species, *R. haasi,* for the remains and considered it intermediate between *R. rattus* and *R. norvegicus* on the basis of cheek-teeth specialization. However, he admitted that it is not especially close to either *R. kiridus* (which has some similarities to *R. edwardsi*), *R. rattus,* or *R. norvegicus,* the three species probably belonging to three separate groups within the genus. The oldest fossil remains from the Middle East excavations are those attributed to *R. haasi* which may date from approximately 120,000 years ago. The Kebara remains (*R. rattus*) were from deposits laid down about 20,000 years ago, while those in the Aba Usba cave date back only 10,000 years. The Aba Usba fossils are the only ones from the Neolithic period when man began settling in villages. However, there is no evidence that these animals were associated with him or shared his permanent living quarters.

The oldest *Rattus rattus* remains in Africa proper are those mentioned by Davis[229] near Livingston, Zambia, a region where no *Rattus* exist today. The fossils were in historical deposits dating from 900 to 1200 A.D.

These discoveries suggest that *Rattus rattus* may have been in the Middle East as early as 8000 B.C. and possibly as much as 10,000 years earlier although the latter is doubtful. Unfortunately the evidence is so fragmentary that a final judgment cannot be made until more complete skeletal remains are unearthed. It seems likely that some of the remains described belong to modern African genera. Nevertheless, such evidence cannot be dismissed out of hand, and the possibility must be entertained that *Rattus rattus* migrated to the Middle East before man is thought to have taken to the seas. Tchernov[226] suggests that *R. rattus* migrated into the Middle East in the upper Levalloiso-Mousterian (more than 20,000 years ago) when the native *Arvicanthis* and *Mastomys* vacated suitable niches for them to move into. This hypothesis lacks direct evidence, but we do know that Indian sailors visited the Malay Archipelago about the beginning of the Christian Era, and that Arab sailors visited India even earlier. The Himyarites of Southern Arabia sent their ships to Africa and southern India early in the first millenium B.C., and it is possible that Pleistocene man had a more elaborate marine technology than so far suspected.[230] *Rattus rattus* doubtless reached the major trading centers of Africa and the Middle East before the Christian Era and was repeatedly introduced into port cities so that a race of seafaring coastal rats developed in close association with man. Eventually they followed him in his travels (primarily by sea, but also overland) around the world, even returning eventually to the major port cities of Malaysia and Indonesia.

Direct proof that *R. rattus* existed in the Mediterranean region in pre-Christian times comes from several sources. The exact identity of the many representations of murids in ancient Greek, Etruscan, and Roman art[231–233] may be questionable,[234] but the reports of skeletons of *Rattus* in ancient archeological sites cannot be ignored. MacArthur[235] was convinced that the rat skeletons found in excavations at Pompeii[234] were those of *R. rattus,* but I have not been able to verify that assumption. These deposites were formed by the eruption of Mount Vesuvius in A.D. 79. Hirst[236] reported what would seem to be conclusive evidence of the existence of *R. rattus alexandrinus* in Egypt during the age of the Pharoahs. Half-digested remains of this species were found in the mummies of sacred birds of ancient Egypt. Koller[234] mentions the same thing, but neither he nor Hirst cite the original reference. Early in the Christian Era rats appear in mosaics and other art objects in the Middle East,[237] suggesting that they were a normal urban inhabitant. Finally, MacArthur[238] cited as proof of the early presence of *R. rattus* in the Western World the discovery of skeletal remains attributed to *R. rattus* in excavations of a neolithic site in the Abu Usba Cave, Mt. Carmel[228] discussed earlier.

The black rat is usually considered to have been introduced into England in the Norman Period. Presumably it returned with the Crusaders from the Middle East to populate Europe and Great Britain. The recent discovery of remains of two *R. rattus* in a Roman well in England dates from before the eighth century and perhaps as early as the fourth century and suggests that the black rat had reached Europe and England at least 400 years before the crusades.[239] MacArthur[240] found many references to rats in early Irish folklore and suggested that they were known there as pests many centuries before the Middle Ages. An old Irish poem contains the lines:

For four vile beasts Ireland hath no fence:
Their bodies lice, their houses rats possess,
Most wicked priests govern their conscience,
And ravening wolves do waste their fields no less.

Many references to rats in 15th and 16th century European literature have led some historians to conclude erroneously that rats had recently arrived there. Barrett-Hamilton and Hinton[183] suggested that rats, almost as soon as they reached Europe were recognized as pests and were immortalized in legend, song, rhyme, and prose; they cite a number of specific references to them in early literature. For example, Chaucer refers to rats in the Pardoner's Tale:

And forth he goth, no lenger wold he tary, into the toun unto a potecary, and praied him that he him wolde sell som poison, that he might his ratouns quell.

Queen Elizabeth's "Acte for preservacon of Grayne" set a price of one penny "for the heades of everie three rattes or twelve myce."

Rosalind, in Shakespeare's *As You Like It* (Act III, Sc. 3) says:

I was never so rhymed that I can remember
Since Pythagoras's time, when I was an Irish rat . . .

"Rhime them to death as they do Irish rats," says Ben Jonson in *Poetaster*.

These latter examples refer to the well-known (at the time) penchant of the Irish for satirizing rats and including them in their poems.

In *Romeo and Juliet* (Act III, Sc. 1), Romeo's friend Mercutio threatens Lady Capulet's nephew, Tybalt, with his sword and says: "Tybalt, you ratcatcher, will you walk?"

Prospero in the *Tempest* (Act I, Sc. 2) says: "The very rats instinctively have quit it."

The antiquity of *R. rattus* in the Middle East and on the east coast of North Africa can no longer be doubted. The Mediterranean Sea probably did not pose any effective barrier to the northward spread of *R. rattus,* which may have become established in southern Europe even before man began using boats. Later, traffic by ship through the Straits of Gibraltar and along the Atlantic coast of Europe and England, perhaps during the Roman conquest of Britain, if not before, undoubtedly contributed to its spread.

b. The History of Plague in the Middle East and Europe

Actual evidence that *R. rattus* was present in pre-Christian and early post-Christian Europe is scarce, and much of it is open to various interpretations. One major historical fact, however, almost demanded its presence in large numbers 500 years after the birth of Christ. This was the great "Justinian Plague" that ravaged the Western civilized world beginning in the 15th year of the reign of Emperior Justinian (A.D. 542) and lasting 50 to 60 years. Before it ended the plague had spread throughout much of Europe and may have claimed as many as 100 million victims. (A "myriad of myriads" are said to have died according to Procopius.[241] This translates to approximately 100 million, a number which Gibbon[241] considers ". . . not wholly inadmissible.") It was clearly the equal of the Black Death 1000 years later.

No one in modern times has seriously questioned the etiology of the Justinian plague. *Yersinia (=Pasteurella) pestis*, the plague bacillus, was doubtless involved. However, it is almost equally certain that not all the major epidemics in the latter half of the sixth century A.D. were caused by *Y. pestis*. Relapsing fever, smallpox, typhoid, malaria—to mention the most devastating—certainly were also pestilential diseases that contributed greatly to the decimation of human populations. I do not wish to deny plague its rightful place in the 6th century pandemic, but merely to place it in proper perspective.

Yersinia pestis is transmitted to man primarily by the Oriental rat flea, *Xenopsylla cheopis,* whose natural host is the rat. In all modern plague epidemics, including the pneumonic form, this flea has been the major, if not the only, primary vector of the disease to men. For plague to have become pandemic and spread rapidly from city to city and country to country, as the Justinian plague is reported to have done for 50 years or more, the rat and its flea must have been well established in the entire Middle East, Mediterranean region, and throughout Europe. Even if we admit that deaths were not all caused by bubonic plague, a good many were, and the inescapable conclusion is that the black rat was widespread throughout the region by the 6th century A.D. Since it could not have appeared overnight, it must have been introduced much earlier.

Repeated references in ancient chronicles to plagues and pestilences attest to their prevalence throughout historical times. In most records the exact nature of the diseases cannot be determined, much less their mode of transmission. The Justinian plague is the first episode of pandemic proportions whose etiology has been established beyond a reasonable doubt, but bubonic plague must have occurred in more localized epidemics even earlier. Hirst[236] believes that the deadly bubonic outbreaks reported by Rufus of Ephesus to have occurred in the Levant about 300 B.C. and in Libya about 50 B.C. were true plague. Biblical accounts of the Philistine plague provide few concrete facts, except that the epidemic was severe. Its exact nature is controversial. Some plagues recorded in the books of Samuel (900 to 200 B.C.) in the Old Testament may well have been bubonic, but opinion is divided, revolving mainly around the exact meaning of the term "emerods" with which the Philistine plague victims were afflicted.

And the hand of the Lord was heavy upon the Azotians, and he destroyed them, and afflicted Azotus and the coasts thereof with emerods. And in the villages and fields in the midst of that country, there came forth a multitude of mice; and there was the confusion of a great mortality in the city.

And he smote the men of every city, both small and great, and they had emerods in their secret parts. And the Gethrites consulted together, and made themselves seats of skins.

The men also that did not die were afflicted with the emerods . . .

Vulgate, I Kings V. 6, 9, 12.
Authorized Version I Samuel 5, 6

"According to the number of the Provinces of the Philistines you shall make five golden emerods, and five golden mice; for the same plague hath been upon you all, and upon your lords. And you shall make the likeness of your emerods, and the likeness of the mice that have destroyed the land, . . "

Vulgate, I Kings VI. 5
Authorized Version I Samuel

These two passages vary somewhat in different translations of the Old Testament, and the actual meaning of key words has never been satisfactorily determined. On only one point is there universal agreement, that a pestilence of some sort occurred during the Israelite struggle for Palestine against the more powerful Philistines, who had settled there after their defeat by Rameses III in Egypt. The "plague" that struck the Philistines (and the Israelites as well) was a severe epidemic, and to it is given the credit for the Hebrew victory over the Philistines, who, in their panic, returned to the Israelites the Ark of the Covenant and thus sealed their fate.

Simpson[242] identified the Philistine plague as bubonic plague and the "emerods" as inguinal buboes. He also interpreted the Biblical passages as meaning that a plague existed *among* "mice" at the same time, instead of the correct meaning that "there was a plague *of* mice." Rodent plagues often accompanied or preceded severe epidemics, but it could seldom be established that the rodent population explosion had any direct etiological relationship to epidemic human disease. MacArthur,[235] to prove the presence of *R. rattus* in pre-Christian Europe, argued (not very convincingly) that the Philistine plague was actually bubonic and inferred that the Philistines were aware of the relationship between the rat and plague, a highly unlikely assumption as it was then believed that plague and pestilence were meted out by the gods. Also, the Biblical account apparently refers to field mice rather than city dwelling rats. Shrewsbury,[243] in a lengthy and less passionate paper, suggested that "emerods" actually meant hemorrhoids and not plague buboes, and that the plague of mice was just that—an outbreak of field mice. Whether the Philistine plague was due to bacillary dysentery, as Shrewsbury suggested, or something else, I do not believe that it was due to *Y. pestis*.

The same or similar arguments apply to the plague that afflicted Sennacherib's army when he invaded Palestine in 701 B.C. It is often assumed that this epidemic was bubonic plague, but little evidence supports this assumption. Shrewsbury[243] did not agree because the black rat, from which the disease was supposed to have spread, had not yet reached Egypt. He preferred to place the first introduction of the black rat into Syria and Palestine somewhat later, coincident with the importation of Indian elephants by Antiochus in his battle against the Jews under Judas Maccabeus (I Maccabees VI, 30 and 37).

The nature of the Philistine plague is important to the present discussion only in that it may provide clues to the presence of *R. rattus* in Palestine as far back as the second millenium B.C. This was a time when the "Sea Peoples," as Mullo Weir[244] called them, were on the move, marching in successive waves across Asia Minor to attack Egypt. Palestine, situated as it is between Asia and Europe to the north and east, and Africa to the south and west was the commercial, political, and military center of the ancient world. It was a hub of both sea and land trading routes and a focus for great movements of people,[245] offering every opportunity for the early distribution of *R. rattus,* which was probably well established in the cities of Palestine when the nomadic Hebrews wrested control of the country from the more civilized worshippers of Baal.

If bubonic plague epidemics occurred before the Christian Era, as they probably did, rats must have been present. In early writings, mice were often associated with plagues and pestilences but never as a direct or even indirect cause. In the Old Testament, plagues were considered to be punishment for arousing God's anger; heavy infestations of mice were merely another sign of His displeasure, as were earthquakes and unusual celestial phenomena. I disagree with MacArthur's[240] interpretation that Biblical writers attributed plague to rats. Mice and rats were not differentiated by the ancients, so we have no way of knowing for certain if the rodents were field mice or *R. rattus*. Many rodent outbreaks associated with famine probably were of field mice, but those associated with plague-like epidemics may well have been of *R. rattus*.

If we accept the substantial evidence that *R. rattus* has been distributed throughout the Middle East, East Africa, and Europe for several thousand years, we must admit that the Oriental rat flea, *Xenopsylla cheopis,* was also present throughout the territory. What about *Rickettsia typhi?* Did it accompany *R. rattus* and *X. cheopis* on their earliest Westward migrations, or was it introduced later; and, if so, when? The answers to these questions may contain the vital clues necessary to determine the origin of *R. prowazekii* and louseborne typhus.

3. History of Early Typhus Epidemics

Dr. Hans Zinsser[246] in *Rats, Lice and History* presented delightfully and with great erudition the history of epidemic typhus. In that magnificent chronicle of one of the most devastating diseases of recorded history, he proposed an intriguing hypothesis of the origin of *R. prowazekii*. He noted that the 16th century Hungarian Wars between Europeans and Turks and their consequences created the circumstances ". . . which gave fleaborne typhus the opportunity of passing from man to man by lice in uninterrupted cycles, short-circuiting the rat-flea phase and adapting the parasitism firmly as a man-louse-man transmission in the form which we now know as the 'classical European type' . . ." No clearer statement of the probable mechanism by which louseborne typhus originated has ever been made. Zinsser also stated in this work the reasons for selecting the 16th century and Hungary as the time and place for the first appearance of *R. prowazekii*. Its evolution from *R. typhi* is beyond reasonable doubt; the accuracy of the exact time and place are perhaps irrelevant to the basic hypothesis that man was directly and intimately involved in its evolution. However, Zinsser based his assumptions on the mistaken belief that *R. rattus* was relatively new to Europe. The popular view at the time was that the Crusaders had introduced it several centuries before, but this was probably in error by at least 1000 years.

Unlike the history of bubonic plague, no clearcut records appear before the Middle Ages referring unquestionably to epidemic typhus in Europe or elsewhere. Most investigators agree that the first recorded epidemic unequivocally due to *R. prowazekii* occurred in 1083 A.D. at the Monastery of La Cava near Salerno, Italy. It was probably not an isolated event since reports of similar epidemics occurring about the same time came from northern Italy and eastern Europe.[247]

MacArthur[248] thought that the great plague of Athens (430 to 428 B.C.) during the Peloponnesian War was due to epidemic typhus and not to smallpox, as Zinsser[246] and others have claimed. MacArthur based his conclusion on Thucydides' detailed and precise description of the epidemic and the clinical syndrome in his history of the Peloponnesian War.[249] Thucydides himself suffered from "plague" and perhaps could write with some authority. The disease apparently was not new to the region since he referred to a similar one having occurred in many places before it appeared in Athens, but of these outbreaks he had no direct knowledge.

Patrick[250] seems to agree with Zinsser that the Athens plague was probably due to smallpox rather than typhus. He cites a description by Diodorus Siculus of another plague in 396 B.C. that befell the Carthaginians during their siege of Syracuse and concurs, again with Zinsser,[246] that it also was due to smallpox. The clinical syndrome differed somewhat from that in the Athens plague, but the rash was similar. However, Thucydides did not mention scarring of those who recovered from the Athens plague. This is a forceful argument against the smallpox theory.

The two most widely discussed candidates for the etiology of the great Athens plague are typhus and smallpox, but the evidence is not conclusive for either. Plague is ruled out mainly because Thucydides did not mention the presence of buboes. Crawfurd[251] considered the possibility of ergotism, but discounted it on clinical grounds. He preferred the diagnosis of typhus, while admitting that the symptoms as described, except for the apparent absence of buboes, could apply equally to plague and typhus. One epidemiological fact that does not support the typhus theory is the dramatic mortality among dogs and other animals that feed on carrion. Crawfurd[251] stated that carrion-feeders could only mean rats, cats, and possibly pigs, none of them lethally susceptible to typhus. Rats, of course, are susceptible to plague; carnivores generally are considerably less likely to contract it.[252] Although dogs and cats have been reported as dying from plague,[253] the evidence was merely presumptive. However, Strong and Teague[254] reported that dogs are susceptible to pneumonic plague. The epizootic among domestic and peridomestic animals in Athens coincided with the human epidemic, but may have been due to something completely unrelated to the human disease.

Despite vigorous support for one or another cause of the great pandemic and highly fatal disease, the most convincing arguments in my opinion have been presented in favor of a disease which in modern times usually has a low mortality. Shrewsbury[255] seriously considered measles as the cause of the Athens plague and Page[256] later concurred. Page stresses that the disease was new to the Athenians, and they were presumably completely without resistance to it. He cites the Fiji epidemic in 1875 in which 25,000 of 100,000 people (all nonimmunes) died of measles in a few months. The course and clinical manifestations of the Fiji epidemic were strikingly similar to those described by Thucydides in Athens over 1400 years earlier.

It is unfortunate that almost everything we know about the Great Plague of Athens has come down to us in the works of one man, Thucydides, writing nearly 2500 years ago. Historical writings, even those of such an astute observer and chronicler, have to be interpreted cautiously. Much early writing about specific events has been done many years after the events took place—sometimes hundreds of years later—such as many books of the Bible.[257] Even Thucydides did not write about the Athens plague or the Peloponnesian Wars until two decades or more thereafter. An elementary rule of historical research that is often ignored is: evidence is valuable in proportion as it dates from a time near the event in question and valueless if it is separated from this event by hundreds of years.[258]

Although there has been a great deal of speculation, we simply do not know for certain the cause or causes of the Athens plague and probably never shall. This is frustrating, for if the epidemic was caused by *R. prowazekii,* a whole new dimension for speculation would be opened. The epidemic is thought to have started in Ethiopia and spread northward to Egypt and Libya, thence over most of western Asia, finally reaching Athens. If it was truly epidemic typhus, this means that typhus is a more ancient disease than Zinsser[246] and others believed, and its place of origin would have to be moved from East Europe to Northeast Africa.

4. Origin and Dispersion of *Rattus norvegicus*

The 16th century typhus epidemics that plagued the adversaries in the Hungarian Wars marked the first well-recorded appearance of the disease in Europe in epidemic form. *Rickettsia prowazekii* may have differentiated from *R. typhi* at that point in history, but it was probably introduced into Eastern Europe from the Middle East as already discussed. An alternative hypothesis, however, is that it originated in southern Russia, east of the Caspian Sea, which is the most likely center of dispersion of the brown rat, *Rattus norvegicus,*[183] where it differentiated from *R. r. caraco,* the Chinese form of the ancestral *R. r. argentiventer* of Indonesia.[215] In modern times *R. norvegicus* has more or less replaced the smaller, less aggressive *R. rattus* in urban centers, particularly in port cities, but considerable controversy persists over when the brown rat began its worldly travels. It apparently followed the trade routes across Central Asia, along which it is still found, into Eastern Europe and the Middle East. As early as the 13th Century A.D. the Mongols under Genghis Khan and his successors built thousands of miles of roads from their mountain retreat at Kara-Korum all the way to Eastern Europe.[261]

According to some writers,[183] *R. norvegicus* did not begin its colonization of the western world until some time during the last two or three centuries. However, remains of *R. norvegicus* have been found recently in 9th and 10th century deposits in Germany[259] indicating its presence in Europe for far longer than previously thought. In any case, its westward migration must have followed that of *R. rattus* by some thousands of years. Once started, *R. norvegicus* rapidly migrated around the world along land and sea routes of human communication. First it populated major ports and trade centers; then, gradually pushing out the weaker less adaptable *R. rattus,* it spread throughout most of the urbanized temperate zone. Primarily a rat of temperate climes it has been less successful in displacing *R. rattus* in tropical regions, except in port cities. According to Lantz[260] *Rattus norvegicus* did not reach the New World until the latter part of the 18th century. It first appeared on the Atlantic Coast, probably brought by ship from England, and did not reach the Pacific Coast until nearly a century later. It followed the black rat, *R. rattus,* by several hundred years.

A reasonably good case might be made for pushing the origin of *R. prowazekii* backward in time and eastward a few thousand miles to Mongolia or even Manchuria. We know nothing of louseborne typhus in China until relatively modern times. Rickettsia prowazekii may have evolved from *R. typhi* somewhere in the vast reaches of northern China, Manchuria, Mongolia, or southeastern Russia by less dramatic processes than those just postulated. The military escapades of the Mongols in the 12th and 13th centuries could have introduced typhus (perhaps a mild form of the disease) into southern Russia, Afghanistan, and Persia, all of which were overrun by the armies of Genghis Khan and his successors.[261]

Still another hypothesis is that *R. prowazekii* originated in association with *R. norvegicus* in southern Russia east of the Caspian Sea. Records of louseborne typhus in this region do not predate the 16th century, but typhus may be endemic as well as epidemic. In the endemic phase, it would attract little attention and could easily have escaped the notice of ancient chroniclers. Fleaborne typhus must surely have existed in *R. norvegicus* in this area, and the presumed close relationship of this rat with man could have led eventually to the selection of a strain of *R. typhi* transmissible from man to man by human lice. The great social upheavals in Eastern Europe during the 16th century provided the conditions for rapid transmission of *R. typhi* through completely nonimmune populations of invading western Europeans; thus more virulent strains of *R. prowazekii* could have emerged and later became established in the Western World.

Well-documented cases of recrudescent typhus in persons of Russian and East European extraction[82,144–149,262,263] may indicate long association of these peoples with typhus. However, the factors predisposing to the development of latent infections are little known. Evidence fails to substantiate that the tendency for parasites to produce latent infections in certain hosts is a function of the length of time they have been associated, but this may be a step in the adaptive processes of accommodation between host and parasite. Western Russia and eastern Europe have long been known to be hyperendemic areas of typhus, and since the Middle Ages have been major centers of dispersal of louseborne typhus to the rest of the world.

5. *Flying Squirrels*

The center of origin of *R. prowazekii* in Eastern Europe or the Middle East may have to be reevaluated in light of the recent isolation of ten strains of rickettsia apparently identical to *R. prowazekii* from flying squirrels (*Glaucomys volans volans*) in Florida and Virginia.[69,264] Specific *R. prowazekii* antibody was found in a high percentage of flying squirrels along the eastern seaboard, but in no other animals tested except one *Peromyscus leucopus*.[265] Extensive biological and biochemical comparison of these strains with human isolates of *R. prowazekii*, rat isolates of *R. typhi*, and tick isolates of *R. canada* have convincingly demonstrated that they are distinct from *R. typhi* and *R. canada* and essentially indistinguishable from human strains of *R. prowazekii*.[20,266] The few differences detected by isoelectric focusing of soluble extracts suggest only slight evolutionary differentiation of the squirrel biotypes from the human strains tested.[20] In these excellent studies, four of the six squirrel strains were tested against two human strains. The squirrel strains were recent isolates in low-egg passage; the human strains were the high-egg passage Breinl strain and its erythromycin-resistant variant. It would be of considerable interest to determine if recent human isolates in low-egg passage are also as similar to the squirrel strains.

Lice (*Neohaematopinus sciuropteri*) and fleas (*Orchopeas howardii*) from flying squirrels in Virginia were naturally infected with *R. prowazekii* and are suspected to be the vectors.[265] Experimental transmission by these arthropods has not been reported as yet. The flea is especially suspect since it will bite man, and eight human cases of serologically confirmed *R. prowazekii* infection were associated with squirrels and other wildlife in the enzootic area.[267]

The origin of *R. prowazekii* in flying squirrels of eastern U.S. is unknown. It seems unlikely that it could have arisen from *R. typhi* independently in Eurasia and north America, and such a postulation may not be necessary. The Sciruidae probably arose in the Palearctic (see Volume I, Chapter 4) and reached North America by migration across the Bering Sea land bridges in the Pleistocene or earlier, perhaps as early as the Oligocene. The initial interchange of placental mammals occurred between Eurasia and North America across Beringia. It is conceivable that they brought the ancestral form of *R. prowazekii* with them, but squirrels are not known to be involved in the ecology of the typhus line in Asia. They are hosts of tick typhus rickettsiae in both Eurasia and North America, and there is little doubt that *R. prowazekii* arose from a tickborne ancestor. However, the data presented earlier overwhelmingly points to the evolution of *R. prowazekii* by adaptation of *R. typhi* (not tickborne rickettsiae directly) to man and the human body louse. The isolation of *R. prowazekii* or related organisms from Eurasian squirrels would force the reevaluation of that hypothesis, but unless or until such isolations are made, the origin of the North American squirrelborne *R. prowazekii* will have to be sought elsewhere.

The squirrel strains of *R. prowazekii* are so similar biologically and antigenically to human strains that an alternate hypothesis readily suggests itself. Louseborne typhus was epidemic in cities along the eastern seaboard in the 18th and 19th centuries. Squirrel fleas feeding on rickettsemic individuals could have transferred the agent back to squirrels and an extrahuman cycle became established. This would neatly explain the presence of *R. prowazekii* in squirrels, but it is pure conjecture at this point. There are too many unanswered questions to be able to explain the facts presently known.

6. The Human Body Louse

A study of the arthropod vector of epidemic typhus, *Pediculus humanus*, provides few clues to the origin of *R. prowazekii*. The sucking lice (Anoplura) are generally host specific, one species infesting a single host species or group of closely related species.[268] The Pediculidae, according to some authors, comprise only three genera: *Pedicinus, Pediculus,* and *Phthirus*. The species of these genera are found only on man and other primates. *Pedicinus* is restricted to Old World monkeys, *Pediculus* and *Phthirus* to man and higher apes.

Ewing[269] divided the genus *Pediculus* into three subgenera: *Pediculus*—varieties of lice infesting man, *Parapediculus*—New World species found on monkeys, *Paenipediculus*—the African species infesting the chimpanzee. The fact that lice closely related to those found on man are also found on New World monkeys and the absence of such closely related lice on apes and monkeys of the Old World prompted Ewing[270] to suggest that the pediculid lice of American monkeys were derived from those on American Indians when the latter reached the tropical regions of the New World in their Southward migrations. Ewing's further work[269] fully supported this hypothesis. Moreover human lice (*Pediculus humanus*) have been found on scalps of pre–Columbian Peruvian Indian mummies.[270]

Only the human louse (*Pediculus*) is biologically and ecologically associated with *R. prowazekii*. Infection of *Pedicinis albidus* fed on monkeys in the acute phase of typhus infection were reported to become infected with *R. prowazekii,*[271] but the only thing actually demonstrated was that the louse could acquire the organism from rickettsemic monkeys. Earlier experiments with *P. longiceps*[272,273] indicated that the organism does not persist in this louse.

The human louse is essentially world-wide in distribution. In some regions, particularly the tropics, the head louse (*P. humanus capitus*) is more common than the body louse; in other areas body lice and head lice have largely been eradicated in recent years. From earliest times man carried his lice wherever he went, yet louseborne typhus, as far as can be determined, developed only during historical times. *Rickettsia prowazekii* has not yet fully adapted to the louse (or vice versa) and the infection it causes is almost invariably fatal. The louse has its own complement of symbiotic microorganisms unrelated to rickettsiae, which it houses in a mycetome and which are hereditarily transmitted from generation to generation. *Pediculus humanus* also transmits *Borrelia recurrentis* (the etiologic agent of relapsing fever), but is not harmed by it.

Unlike *B. recurrentis, R. prowazekii* is apparently an unusual and, for the louse, deadly disease agent picked up from its human host. An early claim by Wilder[274] that he had demonstrated transovarial transmission of *R. prowazekii* from infected female lice to their offspring was based on totally inadequate data as Anderson and Goldberger[275] later showed.

Both *R. prowazekii* and *B. recurrentis* are specific parasites of man and the human body louse transmitted through the feces or by crushing the louse into the bite would; neither is transovarially transmitted; each belongs to a group of related microorganisms that infects other vertebrates and are transmitted by ticks. The major difference between them biologically is that the relapsing fever spirochaete does not kill its insect host, suggesting that the association between louse and spirochaete is more ancient than that between louse and rickettsia.

7. Typhus—A Man-Made Malady?

Rickettsia prowazekii is the only rickettsia in whose evolution man appears to have been intimately involved. We do not know when the first human case of louseborne typhus occurred, but epidemics probably were not a serious problem until the process of urbanization was well under way, and man had begun to congregate in permanent dwellings. Hare[276] is of the opinion that man was not attacked by parasites until he begin to live in villages during the Neolithic period—18,000 to 6,000 B.C. The reference is to typically human parasites; during the preceding Paelolithic period, covering about 98% of man's total existence, he was host to the parasites of other animals.

Whatever the exact time and place of origin of *R. prowazekii,* louseborne typhus might well represent one of the earliest and most devastating examples of what Audy[277] has termed "man-made maladies." Time and time again, inadvertent or deliberate interference in natural host-parasite cycles has resulted in unpleasant consequences, but in few instances has a wild-animal parasite become so specially adapted to man himself. For this to have happened, it was first necessary for a close relationship to have developed between man and *Rattus*. From the earliest days of human history, man, by his very lifestyle, provided an environment in which the rat could flourish. The growing human population and its tendency toward urbanization—clustering together in crowded warrens—radically changed the environment, enormously increasing the habitat suitable for occupation by *Rattus*. This process of urbanization had been going on for thousands of years before the Middle Ages when typhus first appeared in Europe as a recognizable disease. The great civilizations that flourished in the valleys of the Tigris, the Euphrates, and the Nile and along the Mediterranean shores before the Christian Era may or may not have been plagued with typhus. If they were, no reliable record has come down to us that the disease ever reached major epidemic proportions with the possible exception of the Athens plague already mentioned. The rat (probably) and the louse (certainly) had been present as commensals and parasites of Western man for a very long time. We must assume that *R. typhi* was also present in the rats and rat fleas. Sporadic cases of murine typhus probably were common; and under normal circumstances man, louse, rat, flea, and *R. typhi* might have coexisted peacefully without ever seriously upsetting the ecological balance. As long as the rate of transmission of *R. typhi* to man was low and sporadic in nature, there was little chance for the selection of mutant strains suitable for independent transmission between man and louse. This equilibrium, however, was destined to be upset by man himself.

The valleys of the Tigris and Euphrates are usually acknowledged as the cradle of civilization, that is, the beginning of urbanization. Written records begin here with the Sumerians, who migrated into South Babylonia, probably from the east, about 3000 B.C. or earlier. Even before the Sumerians came, there were villages and small urban communities going back to before 5000 B.C.[244] Egypt, another center of early civilization, developed more slowly, but by 3000 B.C. the small agricultural villages and towns became united into a single kingdom under one king. Egyptian recorded history beings with the establishment of the first of 31 Dynasties about 2800 B.C.

With the coming of civilization, tribal battles became more complex and of greater duration. It was not possible to engage in prolonged struggles which were logistically impractical before urbanization and the evolution of religio-political structures. And wage war they did, just as 5000 years later we are yet. Beginning with the 11th Dynasty, the earlier punitive raids on neighboring regions gradually gave way to large scale wars of conquest against Nubia, Libya, Palestine, and Syria. Egypt was conquered and ruled by Semitic and Hurrian invaders from Palestine during the last days of the Middle Kingdom (ca. 1700 to 1600 B.C.). The founding Dynasty of the New Kingdom (ca. 1600 to 1100 B.C.), however, expelled the invaders and pursued them beyond the Euphrates. By the end of the New Kingdom, Egypt had pulled back to the Nile Valley, and a long period of turmoil began during which power and suzerainity changed hands repeatedly until Egypt finally became a Roman province in 30 B.C. The point to be made is that the Near East had been a center of great activity and movement of peoples since the beginning of civilization. The immense importance of this region to the evolution and development of human pathogens cannot be overlooked.

As the human population slowly increased and man became more and more adept at devising defensive structures and offensive weapons, wars became, if not more frequent, at least more prolonged. Armies grew larger and more mobile. Larger and larger areas were involved in disputes. Alexander Severus (The Great King), in the conquest of Mesopotamia (A.D. 233) is said to have fielded an army ". . . of 129,000 horses, clothed in armour of steel; of 700 elephants, with towers filled with archers on their backs; and of 1800 chariots, armed with scythes."[241] The number of foot soldiers is not mentioned, but in any case Gibbon admits that this was probably a gross exaggeration ". . . dictated . . . by the vanity of the monarch, adorned by the unblushing servility of his flatterers, and received without contradiction by a distant and obsequious senate." Nonetheless, armies of several hundred thousand men were often fielded during the Roman Empire.

Disruption of normal civilian activities became more severe and resulting famines more extensive. At some point in early civilization, possibly very early but perhaps not until the Middle Ages, conditions became ripe for the continuous transmission of *R. typhi* from rats and fleas to man and for the selection of a strain that could be transmitted by human lice independent of the rat or the flea. Thus was born *R. prowazekii*.

IV. SUMMARY

Zinsser's hypothesis on the origin of *R. prowazekii* has stood the test of time. The basic assumption is that man is and always has been the major (if not the only) vertebrate host and the human louse the major (if not the only) arthropod host in which *R. prowazekii* differentiated from *R. typhi*. Because neither man nor louse is an ideal host for *R. prowazekii*, the sanctity of this assumption might be questioned, and the ability of either or both to maintain the rickettsia between epidemics is dubious. However, in hyperendemic areas such as Eastern Europe, where typhus occurs sporadically and epidemically, a continuous cycle sufficient to maintain the rickettsiae may exist. The occurrence of a latent form of classical typhus in East Europeans, so-called Brill-Zinsser disease or recrudescent typhus, suggests that man may after all be a suitable host for maintaining *R. prowazekii* during interepidemic periods. Thus there is no compelling reason to alter the hypothesis that *R. prowazekii* originated from *R. typhi* by adaptation to a louse-man cycle in place of a flea-rat cycle. Wisseman et al.[278] have proposed that this may have occurred as a result of the ability of *R. prowazekii* to infect louse gut cells and human nonprofessional phagocytes even when coated with human antityphus antibody.

If our conception is correct that *R. typhi* originated in Asia, the possibility exists that the two organisms arose independently from a tickborne precursor, one in Asia and the other in Africa. This startling idea belies the close antigenic relationship between *R. typhi* and *R. prowazekii*, but it might explain the occurrence of *R. prowazekii* in ticks in Ethiopia. On the other hand, the possibility that *R. rattus, Xenopsylla cheopis*, and *R. typhi* migrated to the Middle East from Southeast Asia tens of thousands of years ago and were already well established when the primitive hunter–gatherers began settling down in permanent dwellings seems much more likely. However, as discussed earlier, it is also possible that the Fertile Crescent is the center of origin not only of *R. prowazekii* but, even before that, of *R. typhi* as well. Perhaps instead of Southeast Asia, the ancestor of *R. typhi* was the tickborne *R. conorii* of the African Savanna or the East European plains, and *R. rattus* only became secondarily involved millenia later. The only thing that seems reasonably clear is that *R. prowazekii* evolved from *R. typhi*. It may never be known if this occurred in Asia, Africa, or eastern Europe.

REFERENCES

1. **McKiel, J. A., Bell, E. J., and Lackman, D. B.,** *Rickettsia canada:* a new member of the typhus group of rickettsiae isolated from *Haemaphysalis leporis-palustris* ticks in Canada, *Can. J. Microbiol.*, 3, 503, 1967.
2. **Murray, E. S. and Gaon, J. A.,** Cross reactions and anamnestic responses between the typhus and Rocky Mountain spotted fever groups of rickettsiae, Rev. 8th Int. Congr. Trop. Med. Mal., Teheran, 1968, 875.
3. **Bengston, I. A.,** Crossing between typhus and spotted fever by complement fixation in 330 sera, *Am. J. Public Health*, 35, 701, 1945.
4. **Hersey, D. F., Colvin, M. C., and Shepard, C. C.,** Studies on the serologic diagnosis of murine typhus and Rocky Mountain spotted fever. II. Human infections, *J. Immunol.*, 79, 409, 1957.
5. **Shepard, C. C., Pedus, M. A., Tzianobos, T., and Warfield, D. T.,** Recent experience with the complement fixation test in the laboratory diagnosis of rickettsial diseases in the United States, *J. Clin. Microbiol.*, 4, 277, 1976.
6. **Plotz, H., Bennett, B. L., Wertman, K., and Snyder, M. J.,** Crossreacting antibodies in Rocky Mountain spotted fever, *Proc. Soc. Exp. Biol. Med.*, 57, 336, 1944.
7. **Parker, R. R., Kohls, G. M., and Steinhaus, E. A.,** Rocky Mountain spotted fever; spontaneous infection in the tick *Amblyomma americanum*, *Public Health Rep.*, 58, 721, 1943.
8. **Philip, R. N., Casper, E. A., Burgdorfer, W., Gerloff, R. K., Hughes, L. E., and Bell, J. E.,** Serologic typing of rickettsiae of the spotted fever group by microimmunofluorescence, *J. Immunol.*, 121, 1961, 1978.
9. **Ormsbee, R., Peacock, M., Philip, R. N., Casper, E., Plorde, J., Gabre-Kidan, T., and Wright, L.,** Serologic diagnosis of epidemic typhus fever, *Am. J. Epidemiol.*, 105, 261, 1977.
10. **Ormsbee, R., Peacock, M., Philip, R., Casper, E., Plorde, J., Gabre-Kidan, T., and Wright, L.,** Antigenic relationships between typhus and spotted fever groups of Rickettsiae, *Am. J. Epidemiol.*, 108, 53, 1978.
11. **Reiss-Gutfreund, R. J., Capucinelli, P., and Cavallo, G.,** The soluble antigens of *Rickettsia prowazekii, R. typhi* and *R. canada*. Investigation of their interrelationship by various serological methods, Z. Immunitaetsforsch., 148, 315, 1975.
12. **Weyer, F. and Reiss-Gutfreund, R. J.,** Verhalten von *Rickettsia montana* und *R. canada* in Kleiderlausen, *Acta Trop.*, 30, 177, 1973.
13. **Weyer, F.,** Extracellular development of rickettsiae, *Acta Virol. Engl. Ed.*, 12, 44, 1968.
14. **Schramek, S.,** Deoxyribonucleic acid base composition of members of the typhus group of rickettsiae, *Acta Virol. Engl. Ed.*, 16, 447, 1972.

15. **Myers, W. F. and Wisseman, C. W., Jr.,** Genetic relatedness among the typhus group of rickettsiae, *Int. J. Syst. Bacteriol.*, 30, 143, 1980.
16. **Tyeryar, F. J., Jr., Weiss, E., Millar, D. B., Bozeman, F. M., and Ormsbee, R. A.,** DNA base composition of rickettsiae, *Science*, 180, 415, 1973.
17. **Obijeski, J. F., Palmer, E. L., and Tzianabos, T.,** Proteins of purified rickettsiae, *Microbios*, 11, 61, 1974.
18. **Eisemann, C. S. and Osterman, J. V.,** Proteins of typhus and spotted fever group rickettsiae, *Infect. Immun.*, 14, 155, 1976.
19. **Osterman, J. V. and Eisemann, C. S.,** Surface proteins of typhus and spotted fever group rickettsiae, *Infect. Immun.*, 21, 866, 1978.
20. **Dasch, G. A., Samms, J. R., and Weiss, E.,** Biochemical characteristics of typhus group rickettsiae with special attention to the *Rickettsia prowazekii* strains isolated from flying squirrels, *Infect. Immun.*, 19, 676, 1978.
21. **Biraud, Y. and Deutschman, S.,** Typhus and typhus-like rickettsia infections. Geographical distribution and epidemiology, *Epidemiol. Rep. League of Nations, Geneva*, 183, 99, 1936.
22. **Mooser, H., Castaneda, M. R., and Zinsser, H.,** The transmission of Mexican typhus from rat to rat by *Polyplax spinulosus*, *J. Exp. Med.*, 54, 567, 1931.
23. **Dyer, R. E., Ceder, E. T., Rumreich, A., and Badger, L. F.,** Experimental transmission of endemic typhus fever of the United States by the rat flea (*Xenopsylla cheopis*), *Public Health Rep.*, 46, 2415, 1931.
24. **Dyer, R. E., Rumreich, A., and Badger, J. E.,** Experimental transmission of endemic typhus with rat fleas, *Xenopsylla cheopis*, *Public Health Rep.*, 46, 1869, 1931.
25. **Lewthwaite, R., Hodgkin, E. P., and Savoor, S. R.,** The typhus group of diseases in Malaya. VI. The search for carriers, *Br. J. Exp. Pathol.*, 17, 309, 1936.
26. **Brigham, G. D. and Dyer, R. E.,** Endemic typhus fever in native rodents, *JAMA*, 110, 180, 1938.
27. **Lillie, R. D., Dyer, R. E., and Topping, N. H.,** Cerebral pathology in rodents in endemic typhus and Rocky Mountain spotted fever, *Public Health Rep.*, 54, 2137, 1939.
28. **Morlan, H. B., Hill, E. L., and Schubert, J. H.,** Serological survey for murine typhus infection in southwest Georgia, *Public Health Rep.*, 65, 57, 1950.
29. **Brigham, G. D.,** A strain of endemic typhus fever isolated from a field mouse, *Public Health Rep.*, 52, 659, 1937.
30. **Sparrow, H.,** Enquere sur la presence du virus typhique chez les souris de Tunis, *Arch. Inst. Pasteur Tunis*, 24, 435, 1935.
31. **Liu, W. T. and Zia, S. H.,** Typhus rickettsia isolated from mice and mouse-fleas during an epidemic in Peiping, *Proc. Soc. Exp. Biol. Med.*, 45, 823, 1940.
32. **Ormsbee, R. A., Hoogstraal, H., Yousser, L. B., Hildebrandt, P., and Attala, W.,** Evidence for extra-human epidemic typhus in the wild animals of Egypt, *J. Hyg. Epidemiol. Microbiol. Immunol.*, 12, 226, 1968.
33. **Marchette, N. J.,** Rickettsioses (tick typhus, Q-fever, Urban typhus) in Malaya, *J. Med. Entomol.*, 2, 339, 1965.
34. **Brown, G. W., Dohany, A. L., Shirai, A., Gan, E., and Huxsoll, D. L.,** Murine typhus in a Malaysian village, *Southeast Asian J. Trop. Med. Public Health*, 8, 99, 1977.
35. **Sankasuwan, V., Pongpradit, P., Bodhidatta, P., Thonglongya, K., and Winter, P. E.,** Murine typhus in Thailand, *Trans. R. Soc. Trop. Med. Hyg.*, 63, 639, 1969.
35a. **Padbidri, V. S., Kulkarni, S. M., Joshi, M. V., Mahadev, P. V. M., and Geevarghese, G.,** Murine rickettsiosis along the western states of Maharashtra State, India, *Indian J. Med. Res.*, 71, 721, 1980.
36. **Heisch, R. B.,** Urban *Rattus* as the main reservoir of murine typhus in Kenya, *J. Trop. Med. Hyg.*, 72, 195, 1969.
37. **Heisch, R. B. and Harvey, A. E. C.,** Rickettsioses in Kenya: Serological reactions of wild rodents and inoculated guinea pigs, *East Afr. Med. J.*, 36, 116, 1959.
38. **Heisch, R. B.,** Rodents as reservoirs of arthorpod-borne diseases in Kenya, *East Afr. Med. J.*, 38, 256, 1961.
39. **Heisch, R. B., Grainger, W. E., Harvey, A. E. C., and Lister, G.,** Feral aspects of rickettsial infections in Kenya, *Trans. R. Soc. Trop. Med. Hyg.*, 56, 272, 1962.
40. **Kalra, S. L. and Rao, K. N. A.,** Typhus group of fevers in Jubbulpore area, *Indian J. Med. Res.*, 37, 373, 1949.
41. **Anastos, G.,** The scutate ticks, or Ixodidae, of Indonesia, *Entomol. Am.*, 30, 1, 1950.
42. **Reiss-Gutfreund, R. J.,** The isolation of *Rickettsia prowazekii* and *mooseri* from unusual sources, *Am. J. Trop. Med. Hyg.*, 15, 943, 1966.
43. **Pang, K. H.,** Isolation of rickettsiae from rat mites during an epidemic in an orphanage, *Proc. Soc. Exp. Biol. Med.*, 48, 266, 1941.

44. **Liu, W. T.,** Isolation of typhus rickettsiae from rat mites, *Liponyssus bacoti,* in Peiping, *Am. J. Hyg.,* 45, 58, 1947.
45. **Lewthwaite, R. and Savoor, S. R.,** The typhus group of fevers in Malaya. VIII. The relation of the tsutsugamushi disease (including rural typhus) to urban typhus, *Br. J. Exp. Pathol.,* 17, 461, 1936.
46. **Brown, G. W., Robinson, D. M., Huxsoll, D. L., Ng, T. S., Lima, K. J., and Sannasey, G.,** Scrub typhus: a common cause of illness in indigenous populations, *Trans. R. Soc. Trop. Med. Hyg.,* 70, 444, 1976.
47. **Rosenthal, T. and Michaeli, D.,** Murine typhus and spotted fever in Israel in the seventies, *Infection,* 5, 82, 1977.
48. **Ito, S., Vinson, J. W., and McGuire, T. J.,** Murine typhus rickettsiae in the oriental rat flea, *Ann. N.Y. Acad. Sci.,* 266, 35, 1975.
49. **Workman, W. G.,** Experimental transmission of endemic typhus fever of the United States by *Xenopsylla astia, Public Health Rep.,* 48, 795, 1933.
50. **Blanc, G. and Baltazard, M.,** Transmission experimentale du typhus murin par la puce de l'homme *(Pulex irritans), C. R. Seances Soc. Biol., Paris,* 124, 1058, 1937.
51. **Dove, W. E. and Shelmire, B.,** Tropical rat mites, *Liponyssus bacoti* Hirst, vectors of endemic typhus, *JAMA,* 97, 1506, 1931.
52. **Dove, W. E. and Shelmire, B.,** Some observations on tropical rat mites and endemic typhus, *J. Parasitol.,* 18, 159, 1932.
53. **Blanc, G. and Baltazard, M.,** La contamination par voie muqueuse, mecanisme habituel de transmission du typhus murine dans la nature. Role du virus sec des dejections d'ectoparasites dans l'epidemiologie de typhus, *Bull. Acad. Natl. Med. Paris,* 120, 109, 1938.
54. **Violle, H.,** Contribution a l'etude des typhus exanthematique endemiques, *C.R. Acad. Sci.,* 206, 382, 1938.
55. **Worth, C. B. and Rickard, E. R.,** Transmission of murine typhus in roof rats in the absence of ectoparasites, *Am. J. Trop. Med.,* 31, 301, 1951.
56. **Rickard, E. R.,** Attempted transmission of murine typhus to roof rats by ingestion of food heavily contaminated with infected flea feces, *Am. J. Trop. Med.,* 31, 311, 1951.
57. **Nicolle, C., Giroud, P., and Sparrow, H.,** Presence exceptionelle du virus typhique murin dans les urines des rats infectes experimentalement par a virus, *Arch. Inst. Pasteur Tunis,* 23, 1, 1934.
58. **Le Chuiton, F., Pirot, R., Berge, C., and Pennaneac'h J.,** La voie digestive dans la transmission du typhus murin a bord des batiments de guerre, *Bull. Acad. Nat. Med., Paris,* 119, 175, 1938.
59. **Beck, M. D. and Van Allen, A.,** Typhus fever in California, 1916–1945, inclusive. An epidemiologic and field laboratory study, *Am. J. Hyg.,* 45, 335, 1947.
60. **Brigham, G. D.,** Two strains of endemic typhus fever virus isolated from naturally infected chicken fleas *(Echidnophaga gallinacea), Public Health Rep.,* 56, 1803, 1941.
61. **Alicata, J. E.,** Experimental transmission of endemic typhus fever by the stick tight flea, *Echidnophaga gallinacea, J. Wash. Acad. Sci.,* 32, 57, 1942.
62. **Clifford, C. M., Sonenshine, D. E., Atwood, E. L., Robbins, C. S., and Hughes, L. E.,** Tests on ticks taken from wild birds collected in the eastern United States for rickettsiae and viruses, *Am. J. Trop. Med. Hyg.,* 18, 1057, 1969.
63. **Bozeman, F. M., Elisberg, B. L., Humphries, J. W., Runcik, K., and Palmer, D. B., Jr.,** Serologic evidence of *Rickettsia canada* infection in man, *J. Infect. Dis.,* 121, 367, 1970.
64. **Lane, R. S., Philip, R. N., and Casper, E. A.,** Ecology of tickborne agents in California. II. Further observations on rickettsiae, in *Proc. Rocky Mt. Lab. Conf. Rickettsiae Rickettsial Dis.,* Academic Press, New York, 1980, in press.
65. **Hamidi, A. N., Saadatezadeh, H., Tarasevich, I. V., Arata, A. A., and Fathangazad, H.,** A serological study of rickettsial infections in Iranian small mammals, *Bull. Soc. Pathol. Exot.,* 67, 607, 1974.
66. **Gispen, R.,** The role of rats and mites in the transmission of epidemic scrub typhus in Batavia, *Doc. Neerl. Indones. Morbis Trop.,* 2, 23, 1950.
67. **Reiss-Gutfreund, R. J.,** Isolement de souches de *Rickettsia prowazeki* a partir du sang des animaux domestiques d'Ethiopie et de leurs tiques, *Bull. Soc. Pathol. Exot.,* 48, 602, 1955.
68. **Reiss-Gutfreund, R.,** Un nouveau reservoir de virus pour *Rickettsia prowazeki:* les animaux domestiques et leurs tiques, *Bull. Soc. Pathol. Exot.,* 49, 946, 1956.
69. **Bozeman, F. M., Masiello, S. A., Williams, M. S., and Elisberg, B. L.,** Epidemic typhus rickettsiae isolated from flying squirrels, *Nature (London),* 255, 545, 1975.
70. **Bozeman, F. M., Williams, M. S., Stocks, N. I., Chadwick, D. P., Elisberg, B. L., Sonenshine, D. E., and Lauer, D. M.,** Ecologic studies on epidemic typhus infection in the eastern flying squirrel, in *2nd Int. Sypm. Rickettsiae Rickettsial Dis.,* Kazar, J., Ormsbee, R. A., Tarasevich, I. N., Eds., Slovak Academy of Sciences, Bratislava, Czechoslovakia, 1978, 493.

71. **Burgdorfer, W.**, Observations on *Rickettsia canada*, a recently described member of the typhus group rickettsiae, *J. Hgy. Epidemiol. Microbiol. Immunol.*, 12, 26, 1968.
72. **Brinton, L. and Burgdorfer, W.**, Fine structure of *Rickettsia canada* in tissues of *Dermacentor andersoni* Stiles, *J. Bacteriol.*, 105, 1149, 1971.
73. **Ignatovich, V. F. and Grokhovskaya, I. M.**, Relationship between Ixodoidea ticks and *Rickettsia canada*, *Med. Parazitol. Parazit. Bolenzi*, 45, 313, 1976.
74. **Burgdorfer, W. and Brinton, L. P.**, Intranuclear growth of *Rickettsia canada*, a member of the typhus group, *Infect. Immun.*, 2, 112, 1970.
75. **Nicolle, C., Comte, C., and Conseil, E.**, Transmission du typhus exanthematique par le pou du corps, *C.R. Acad. Sci., Paris*, 149, 486, 1909.
76. **Nicolle, C., Comte, L., and Conseil, E.**, Recherches experimentales. IV. Transmission experimentales du typhus exanthematique par le pou du corps, *Ann. Inst. Pasteur Paris*, 24, 261, 1910.
77. **Reiss-Gutfreund, R. J.**, Nouveau isolements de *Rickettsia prowazeki* a partir d'animaux domestiques et de tiques, *Bull. Soc. Pathol. Exot.*, 54, 284, 1961.
78. **Tarizzo, M. L.**, Geographic distribution of louse-borne diseases, in The Control of Lice and Louse-borne Diseases, *Pan Am. Health Organ. Sci. Publ.*, 263, 50, 1973.
79. **Wisseman, C. L., Jr.**, Geographic distribution of louse-borne diseases. Discussion, in The Control of Lice and Louse-borne Diseases, *Pan. Am. Health Organ. Sci. Publ.*, 263, 60, 1973.
80. **Murray, E. S. and Torrey, S. B.**, Virulence of *Rickettsia prowazeki* for head lice, *Ann. N.Y. Acad. Sci.*, 266, 25, 1975.
81. **Weyer, F.**, Rickettsiae and lice, in *Proc. 2nd Int. Symp. Rickettsiae Rickettsial Dis.*, Kazar, J., Ormsbee, R. A. and Tarasevich, J. N., Eds., Slovak Academy of Sciences, Bratislava, Czechoslovakia, 1978, 515.
82. **Zinsser, H.**, Varieties of typhus virus and the epidemiology of the American form of European typhus fever (Brill's disease), *Am. J. Hyg.*, 20, 513, 1934.
83. **Giroud, P., Le Gac, P., Brizard, H., and Laurent, C.**, Comportement des serums de divers animaux domestiques, de l'Oubangui-Chari, vis-a-vis de l'antigene epidemique, *Bull. Soc. Pathol. Exot.*, 45, 313, 1952.
84. **Giroud, P., Pfister, R., Ridet, J., and Roger, F.**, Ce Que l'on peut conclure de constations serologiques faites vis-a-vis des rickettsioses sur des Africains et des animaux domestiques en Haute-Volta (A.E.F.), *Bull. Soc. Pathol. Exot.*, 46, 650, 1953.
85. **Giroud, P. and Jadin, J.**, Comportement des animaux domestiques au Ruanda-Urundi (Congo-Belge) vis-a-vis l'antigene epidemique, *Bull. Soc. Pathol. Exot.*, 46, 870, 1953.
86. **Giroud, P. and Giroud, M. L.**, Agglutination des rickettsies, test de sero-protection et reaction d'hypersensibilite cutanee, *Bull. Soc. Pathol. Exot.*, 37, 84, 1944.
87. **Giroud, P. and Le Gac. P.**, Parente serologique entre la fievre boutonneuse et le typhus epidemique, *C.R. Seances Soc. Biol. Paris*, 142, 436, 1948.
88. **Zarate Aquino, M. L. and Varela, G.**, Distinction del tifo clasico del murino por la microagglutinacion de Giroud, *Med. Rev. Mexico*, 44, 297, 1964.
89. **Fiset, P., Ormsbee, R. A., Silberman, R., Peacock, M., and Spielman, S. H.**, A microagglutination technique for detection and measurement of rickettsial antibodies, *Acta Virol. Engl. Ed.*, 13, 60, 1969.
90. **Iman, I. Z. E. and Alfy, L.**, Complemtnt-fixing antibodies against epidemic and murine typhus in domestic animals in U.A.R., *J. Egypt. Public Health Assoc.*, 38, 101, 1963.
91. **Iman, I. Z. E. and Alfy, L.**, Evidence of typhus infection in domestic animals in Egypt, *Bull. W.H.O.*, 35, 123, 1966.
92. **Ormsbee, R., Burgdorfer, W., Peacock, M., and Hildebrandt, P.**, Experimental infections of *Rickettsia prowazeki* among domestic livestock and ticks, *Am. J. Trop. Med. Hyg.*, 20, 117, 1971.
93. **Philip, C. B. and Iman, I. Z. E.**, New concepts on the epidemiology of typhus fever, in 1st Int. Conf. Vaccines Against Viral Rickettsial Dis. of Man, Pan Am. Health Organ., W.H.O., Washington, D.C., 517, 1967.
94. **Iman, I. Z. E.**, Domestic animals and their role in epidemic typhus, *J. Egypt. Public Health Assoc.*, 43(Suppl.), 165, 1968.
95. **Philip, C. B., Hoogstraal, H., Reiss-Gutfreund, R. J., and Clifford, C. M.**, Evidence of rickettsial disease agents in ticks from Ethiopian cattle, *Bull. W.H.O.*, 35, 127, 1966.
96. **Burgdorfer, W., Ormsbee, R. A., Schmidt, M. L., and Hoogstraal, H.**, A search for the epidemic typhus agent in Ethiopian ticks, *Bull. W.H.O.*, 48, 563, 1973.
97. **Philip, C. B., Lackman, D. B., Bell, E. J., and Hughes, L. E.**, Laboratory identification of typhus isolated by Reiss-Gutfreund from Ethiopian livestock ticks, *Am. J. Trop. Med. Hyg.*, 15, 950, 1966.
98. **Burgdorfer, W. and Ormsbee, R. A.**, Development of *Rickettsia prowazeki* in certain species of ixodid ticks, *Acta Virol. Engl. Ed.*, 12, 36, 1968.

99. **McDade, J. E. and Wisseman, C. J., Jr.**, Studies of proposed extrahuman reservoirs (ticks, including species of *Hyalomma* and *Boophilus*) of epidemic typhus caused by *Rickettsia prowazekii*, *Pan Am. Health Organ. Sci. Publ.*, 263, 113, 1973.
100. **Reiss-Gutfreund, R. J.**, The epidemiology of rickettsioses on the Ethiopian high plateau. A six months survey from October 1964 to April 1965, *Am. J. Trop. Med. Hyg.*, 16, 186, 1967.
101. **Dolgov, G. F.**, Study of possible circulation of *Rickettsia prowazeki* in nature, *Rev. 8th Int. Cong. Trop. Med. Malar., Teheran* (Abstr.), 868, 1968.
102. **Dolgov, G. F., Dutova, G. M., Balayeva, N. M., V'yukov, V. N., and Zhmayeva, Z. M.**, Possible reservoirs of *Rickettsia prowazeki* in nature, *Zh. Mikrobiol. Epidemiol. Immunobiol.*, 45, 150, 1968.
103. **Dolgov, G. F. and Dutova, G. M.**, Search of *Rickettsia prowazeki* extra-human reservoir, *Proc. XIII Int. Congr. Entomol. Leningrad*, 3, 149, 1972.
104. **Philip, C. B., Lackman, D. B., Philip, R. N., Schenone, H., and Coscaron, S.**, Serological evidence of rickettsial zoonoses in South American domestic animals, *Acta Med. Biol. Niigata*, 15 (Suppl.), 53, 1967.
105. **Varela, G. and Velasco, R.**, Nuevos aspectos de la epidemiologia del tifo exploracion serologica de animales domesticos, *Rev. Inst. Salubr. Enferm. Trop. Mexico City*, 25, 171, 1965.
106. **Schmatz, H. D., Krauss, H., Viertel, P., Ismail, A. S., and Hussein, A. A.**, Seroepidemiological investigations in domestic ruminants from Egypt, Somalia and Jordan for the demonstration of complement fixing antibodies against rickettsia and chlamydia, *Acta Trop.*, 35, 101, 1978.
107. **Hoogstraal, H., Kaiser, M. N., Ormsbee, R. A., Osborn, D. J., Helmy, I., and Gaber, S.**, *Hyalomma* (*Hyalommina*) *rhipicephaloides* Neumann (Ixodoidea: Ixodidae): its identity, hosts and ecology, and *Rickettsia conori, R. prowazeki* and *Coxiella burneti* infections in rodent hosts in Egypt, *J. Med. Entomol.*, 4,391, 1967.
108. **Varela, G., Mariotte, C. O., and Silva, R.**, Complement fixation in rat blood sera, *Science*, 104, 303, 1946.
109. **Juminer, B.**, *Rattus norvegicus* et rickettsioses. Role du rat en tant que reservoir de virus probable du typhus epidemique et de la fievre Q, *Arch Inst. Pasteur Tunis*, 36, 173, 1959.
110. **Mosing, H.**, Le Typhus exanthematique en Pologne, *Bull. Off. Int. Hyg. Publique*, 30, 1715, 1938.
111. **Reiss-Gutfreund, R. J. and Croce, F.**, The susceptibility of lambs for the experimental infection with *Rickettsia prowazeki, G. Batteriol. Virol. Immunol. Ann. Osp. Maria Vittoria Torino*, 64, 144, 1971.
112. **Sarycheva, N. I. and Chirov, P. A.**, Experimental infection of domestic animals with *Rickettsia prowazeki* and *R. canada, Zh. Mikrobiol. Immunobiol.*, 9, 101, 1976.
113. **Philip, C. B., Hughes, L. E., Lackman, D. B., and Bell, E. J.**, Susceptibility of certain domestic animals to experimental infection with *Rickettsia prowazeki, Am. J. Trop. Med. Hyg.*, 16, 758, 1967.
114. **Ormsbee, R. A., Peacock, M. G., Bell, E. J., and Burgdorfer, W.**, Experimental infections of lambs with *Rickettsia prowazeki, Am. J. Trop. Med. Hyg.*, 20, 950, 1971.
115. **Giroud, P.**, Au sujet de l'antigene isole a Addis Abeba dans le sang de la chevre et sur tique, *Bull. Soc. Pathol. Exot.*, 54, 698, 1961.
116. **Grokhovskaya, I. M., Ignatovich, V. F., and Sidorov, V. E.**, Susceptibility of ticks of the superfamily Ixodoidea to *Rickettsia prowazeki, Med. Parasitol., Moscow*, 35, 199. 1966.
117. **Grokhovskaya, I. M., Ignatovich, V. F., and Sidorov, V. E.**, Ixodoidea ticks and *Rickettsia prowazeki*, in *Biological Interrelationships of Blood Sucking Arthropods with the Agents of Human Diseases*, Petrishcheva, P. A., Ed., Akademiya Meditsinskikh Nauk SSSR, Moscow, 1967, 126.
118. **Klimentova, A. A. and Perfil'yev, P. P.**, Punaises, puces et tiques comme transporteurs du virus du typhus exanthematique dans les conditions experimentales, *Sb. Rab. Posvyashch. Dvadtsatip. Nauchn. Deyat. Professora Evgeniya Nikanorovicha Pavlovskogo 1909–1934 Izd. Vses. Inst. Eksp. Med. Moskva Leningrad*, 71, 1935; as cited in **Grokhovskaya, I. M., Ignatovicy, V. F., and Sidorov, V. E.**, *Biological Interrelationships of Bloodsucking Arthropods with the Agents of Human Diseases*, Petrishcheva, P. A., Ed., Akademiya Meditsinskikh Nauk SSSR, Moscow, 1967, 126.
119. **Ignatovich, V. F. and Grokhovskaya, I. M.**, Study of possible routes of *Rickettsia prowazeki* transmission by ticks of the superfamily Ixodoidea, *Med. Parazitol. Parazit. Bolezni*, 37, 708. 1968.
120. **Grokhovskaya, I. M., Sidorov, V. P., Kryuchechnikov, V. N., and Ignatovich, V. F.**, Comparison of interrelationship between blood sucking arthorpods and *Rickettsia prowazeki, Rev. 8th Int. Cong. Trop. Med. Malar.*, 866, 1968.
121. **Burgdorfer, W., Ormsbee, R. A., and Hoogstraal, H.**, Ticks as vectors of *Rickettsia-prowazeki*—a controversial issue, *Am. J. Trop. Med. Hyg.*, 21, 989, 1972.
122. **Zinsser, H. and Casteneda, M. R.**, Studies on typhus fever. VIII. Ticks as possible vectors of the disease from animals to men, *J. Exp. Med.*, 44, 11, 1931.

123. **Weyer, F.**, Die Kunstliche Infektion von Zecken mit Rickettsien und anderen Krankheitserregern, *Zentralbl. Bakteriol. Parasitenkd. Infektionskr. Hyg. Abt. 1 Orig.*, 152, 449, 1948.
124. **Ignatovich, V. E. and Grokhovskaya, I. M.**, Experimental prerequisites for substantiating the possible role of ticks in the epidemiology of typhus fever, *Vestn. Akad. Med. Nauk SSSR*, 25, 22, 1970.
125. **Kesarev, I. P. and Prodan, Z. G.**, Experiments on parenteral infection of argasid ticks, *Ornithodoros papillipes* by *Rickettsia prowazeki, Probl. Parazitol.*, 2, 61, 1963.
126. **Kordova, N. and Rehacek, J.**, Microscopical investigations of organs from ticks infected with *Rickettsia prowazeki, Acta Virol. (Engl. Ed.)*, 8, 465, 1964.
127. **Piontkovskaya, S. P. and Korshunova, O. S.**, Experiments of infection in *Dermacentor marginatus* Sulz. and *D. pictus* Herm. ticks with *Rickettsia prowazeki, Vopr. Kraj. Obshchestvo Exp. Parasitol.*, 8, 29, 1953 (In Russian); as cited in **Burgdorfer, W. and Varma, M. G. R.**, *Annu. Rev. Entomol.*, 12, 347, 1967.
127a. **Burgdorfer, W. and Varma, M. G. R.**, Trans-stadial and transovarial development of disease agents in arthropods, *Annu. Rev. Entomol.*, 12, 347, 1967.
128. **Rehacek, J., Brezina, R., and Majerska, M.**, Multiplication of rickettsiae in tick cells in vitro, *Acta Virol. (Engl. Ed.)*, 12, 41, 1968.
129. **Giroud, P., Capponi, M., and Rageau, J.**, Le typhus epidemique peut-il se conserver en dehors du pou et de l'homme son veritable reservoir? *Bull. W.H.O.*, 35, 119, 1966.
130. **Reiss-Gutfreund, R. J.**, Some speculations concerning extrahuman *Rickettsia mooseri, J. Hyg. Epidem. Microbiol. Immunol.*, 12, 133, 1968.
131. **Sarycheva, N. I.**, The possible circulation of *Rickettsia prowazeki* among animals (according to the data of serological studies), *Zh. Mikrobiol.*, 10, 125, 1976.
132. **Pecic, J. and Borjanovic, S.**, Bedeutung des funder Komplementbindender Antikorper fur *Rickettsia prowazeki* bei Haustieren en endemischen Gebieten des Flecktyphus in Serbien, in *2nd Int. Arbeitskolloquium uber Naturherde von Infektionskrankheiten in Zentraleuropa*, Sixe, W. and Troger, H., Eds., Biologisches Forschungsinstitut, Limburg, West Germany, 1976, 341.
133. **Giroud, P. and Giroud, M. L.**, Pouvoir pathogene et antigene du poumon de chien adult infecte par voie tracheale avec le virus du typhus historique, *C.R. Seances Soc. Biol. Paris*, 138, 476, 1944.
134. **Combiescu, D., Zotta, G., Manciulescu, E., Pop, A., and Tascau, J.**, Die Schutzinpfung des Meerschwein chens gegen Klassischen Flecktyphus mit Hilfe formolgetoteter Rickettsein aus Mause und Hundelunge, *Z. Hyg. Infektionskr.*, 123, 612, 1942.
135. **Giroud, P.**, Infection du poumon de mouton avec le virus du typhus historique, *C.R. Seances Soc. Biol. Paris*, 139, 280, 1945.
136. **Sergent, E. and Horrenberger, R.**, Utilization de la chevre pour la preparation du vaccine non vivant contre le typhus exanthematique avec du virus provenant de pneumonie rickettseinne provoquee, *Arch. Inst. Pasteur Alger.*, 22, 8, 1944.
137. **Horrenberger, R. and Renoux, G.**, Utilisation du mouton pour la preparation du vaccin antityphique (antirickettsien) non vivant d'apres la methode de Durand et Giroud, *Arch. Inst. Pasteur Alger.*, 21, 4, 1943.
138. **Ormsbee, R. A.**, The hypothesis of extrahuman reservoir of *Rickettsia prowazeki, Pan Am. Health Organ. Sci. Publ.*, 263, 104, 1973.
139. **Burgdorfer, W.**, Possible extrahuman reservoir of *Rickettsia prowazeki, Pan Am. Health Organ. Sci. Publ.*, 263, 109, 1973.
140. **Philip, C. B.**, A review of growing evidence that domestic animals may be involved in cycles of rickettsial zoonoses, *Zentralbl. Bakteriol., Parasitenkd. Infektionskr. Hyg., Abt. 1 Orig.*, 206, 343, 1968.
141. **Weyer, F.**, Progress in ecology and epidemiology of rickettsioses, *Acta Trop.*, 35, 5, 1978.
142. **Murray, E. S., Psorn, T., Djakovic, P., Sielski, S., Broz, V., Ljupsa, F., Gaon, J., Pavlevic, R., and Snyder, J. C.**, Brill's disease. IV. Study of 26 cases in Yugoslavia, *Am. J. Public Health*, 41, 1359, 1951.
143. **Gaon, A. J. and Murray, S. E.**, The natural history of recrudescent typhus (Brill-Zinsser) disease in Bosnia, *Bull. W.H.O.*, 35, 133, 1966.
144. **Milosovicova, A., Tarabcak, M., and Wernisch, Z.**, Incidence of Brill-Zinsser disease in East Slovakia in 1956–1967, *J. Hyg. Epidem. Microbiol. Immunol.*, 14, 393, 1970.
145. **Gaon, J. A.**, The role of Brill-Zinsser disease as a source of new cases of typhus fever, *Folia Microbiol. (Prague)*, 21, 507, 1977.
146. **Pecic, J., Boskovic, R., Bojanovic, S., and Keemonovic, M.**, Brill-Zinsser'sche Krankheit in Servien von 1964 bis 1975 (Kurzfassung), *Tropenmed. Parasitol.*, 28, 283, 1977.
147. **Mittermayer, T.**, Morbus Brill-Zinsser in der Slowakei, *Infection*, 5, 76, 1977.
148. **Buzdugan, I.**, Some aspects concerning the Brill-Zinsser (B.Z.) disease appearance, *Folia Microbiol. (Prague)*, 21, 507, 1977.

149. **Boskovic, R., Pecic, J., Vukovic, B., and Borjanovic, S.,** Restantikorper gegen *Rickettsia prowazeki* bei einen Aussiedlergruppe aus ehemaligen endemischen Gebieten (Kurzfassung), *Tropenmed. Parasitol.*, 28, 282, 1977.
150. **Murray, E. S. and Gaon, J. A.,** Incidence of *Rickettsia prowazeki* infections in an endemic focus of louse-borne typhus: factors influencing the occurrence of epidemics, in The control of lice and louse-borne diseases, *Pan Am. Health Organ. Sci. Publ.*, 263, 66, 1973.
151. **Murray, E. S. and Snyder, J. C.,** Brill-Zinsser disease: the interepidemic reservoir of epidemic louse-borne typhus fever, *8th Int. Congr. Microbiol. Rome*, 4, 31, 1953.
152. **Gaon, A. J.,** Brill-Zinsser Disease in Bosnia (Yugoslavia) and factors on which depends its role as a reservoir of infection, in *Proc. 2nd Int. Symp. Rickettsiae Rickettsial Dis.*, Kazar, J., Ormsbee, R. A., and Tarasevich, J. N., Eds., Slovak Academy of Sciences, Bratislava, Czechoslovakia, 449, 1978.
153. **Weil, E. and Felix, A.,** Zur serologischen Diagnose des Fleckfiebers, *Wien. Klin. Wochenschr.*, 29, 33, 1916.
154. **Weil, E. and Felix, A.,** Ueber die Beziehungen der Fleckfieberagglutination zum Fleckfiebererreger, *Z. Immunitaetsforsch. Exp. Therap.*, 31, 457, 1921.
155. **Felix, A.,** Serological types of typhus and corresponding types of *Proteus*, *Trans. R. Soc. Trop. Med. Hyg.*, 27, 147, 1933.
156. **Castaneda, M. R.,** Studies on the mechanism of immunity in typhus fever. Complement fixation test in typhus fever, *J. Immunol.*, 31, 285, 1936.
157. **Maxcy, K. F.,** An epidemiological study of endemic typhus (Brill's disease) in the southeastern United States, *Public Health Rep.*, 41, 2967, 1926.
158. **Maxcy, K. F.,** Typhus fever in the United States, *Public Health Rep.*, 44, 1735, 1929.
159. **Mooser, H. and Dummer, C.,** Experimental transmission of endemic typhus of the southeastern Atlantic states by the body louse, *J. Infect. Dis.*, 46, 170, 1930.
160. **Sinclair, C. G. and Maxcy, K. F.,** Mild typhus (Brill's disease) in the lower Rio Grande Valley, *Public Health Rep.*, 40, 241, 1925.
161. **Dyer, R. E., Rumreich, A., and Badger, L. F.,** Typhus fever, a virus of the typhus type derived from fleas collected from wild rats, *Public Health Rep.*, 46, 334, 1931.
162. **Mooser, H., Castaneda, M. R., and Zinsser, H.,** Rats as carriers of Mexican typhus fever, *JAMA*, 97, 231, 1931.
163. **Mooser, H.,** Essai sur l'historie naturelle du typhus exanthematique, *Arch. Inst. Pasteur Tunis*, 21, 1, 1932.
164. **Mooser, H.,** Tabardillo, an American variety of typhus, *J. Infect. Dis.*, 44, 186, 1929.
165. **Mooser, H. and Castaneda, M. R.,** The multiplication of the virus of Mexican typhus fever in fleas, *J. Exp. Med.*, 55, 307, 1932.
166. **Nicolle, C.,** L'Immunite et l'immunization contre le typhus exanthematique, *Rev. Immunol.*, 1, 9, 1935.
167. **Mooser, H.,** Estudio bio-quimico de la reaccion de Weil-Felix. Communiacion preliminair, *Gac. Med. Mex.*, 66, 59, 1935.
168. **Baker, A. C.,** The typical epidemic series, *Am. J. Trop. Med.*, 23, 559, 1943.
169. **Zinsser, H.,** The rickettsia diseases. Varieties, epidemiology and geographical distribution, *Am. J. Hyg.*, 25, 430, 1937.
170. **Gordon, J. E.,** The clinical features of rickettsial diseases, in *Virus and Rickettsial Diseases*, Harvard University Press, Cambridge, Mass., 1940, 828.
171. **Wolbach, S. B.,** The rickettsial diseases. A general survey, in *Virus and Rickettsial Diseases*, Harvard University Press, Cambridge, Mass., 1940, 789.
172. **Sambon, L. W.,** The parasitic acarines of animals and the part they play in the causation of the eruptive fevers and other diseases of man. Preliminary considerations based upon an ecological study of typhus fever, *Ann. Trop. Med. Parasitol.*, 22, 67, 1928.
173. **Cooper, D. B.,** *Epidemic Diseases in Mexico City, 1761–1813*, University of Texas Press, Austin, 1965, 1.
174. **Nicolle, C.,** Unite ou pluralite de typhus, *Bull. Soc. Pathol. Exot.*, 26, 316, 1933.
175. **Heyerdahl, T.,** *Kon Tiki*, Rand McNally, Chicago, 1950, 1.
176. **Voss, J. C.,** *The Adventuresome Voyages of Captain Voss*, Charles E. Lauriat Co., Boston, 1930, 1.
177. **Heyerdahl, T.,** *American Indians in the Pacific*, George Allen and Unwin Ltd., London, 1952, 1.
178. **Vavilov, N. I.,** The origin, variation, immunity and breeding of cultivated plants (trans. K. Starr Chester), *Chron. Bot.*, 13, 1, 1949–1950.
179. **Barrau, J.,** Ed., Introduction, in *Plants and the Migration of Pacific Peoples*, Bishop Museum Press, Honolulu, 1963, 1.

180. **Carter, G. F.,** Movement of people and ideas across the Pacific, in *Plants and the Migration of Pacific Peoples,* Barrau, J., Ed., Bishop Museum Press, Honolulu, 1963, 7.
181. **Ferdon, E. N., Jr.,** Polynesian origins, *Science,* 141, 499, 1963.
182. **Donaldson, H. H.,** The history and zoological position of the albino rat, *J. Acad. Nat. Sci. Philadelphia,* 15 (2nd Ser.), 365, 1912.
183. **Barrett-Hamilton, G. E. H. and Hinton, M. A. C.,** *A History of British Mammals,* Part 18, Gurney and Jackson, London, 1916, 575.
184. **Strandtman, R. W. and Wharton, G. W.,** *A Manual of Mesostigmatid Mites Parasitic on Vertebrates,* The Institute of Acarology, College Park, Md., 1958, 1.
185. **Huff, G. C.,** Studies on the evolution of some disease-producing organisms, *Rev. Biol.,* 13, 196, 1938.
186. **Mooser, H., Varelo, G., and Pilz, H.,** Experiments on the conversion of typhus strains, *J. Exp. Med.,* 59, 137, 1934.
187. **Worth, C. B. and Rickard, E. R.,** Evaluation of the efficiency of common cotton rat ectoparasites in the transmission of murine typhus, *Am. J. Trop. Med. Hyg.,* 31, 295, 1951.
188. **Zemskaya, A. A.,** Gamasid mites (Gamasoidea) as vectors of disease causative agents, *Zool. Zh.,* 46, 1771, 1999.
189. **Burgdorfer, W.** Investigation of "transovarial transmission" of *Rickettsia rickettsii* in the wood tick *Dermacentor andersoni, Exp. Parasitol.,* 14, 152, 1963.
190. **Furman, D. P. and Radovsky, F. J.,** A new species of *Ornithonyssus* from the white-tailed antelope squirrel, with a rediagnosis of the genus *Ornithonyssus, Pan Pac. Entomol.,* 39, 75, 1963.
191. **Rosen, L., Tesh, R. B., Line, J. C., and Cross, J. H.,** Transovarial transmission of Japanese encephalitis virus by mosquitoes, *Science,* 199, 909, 1978.
192. **Tesh, R. B. and Gubler, D. J.,** Laboratory studies of transovarial transmission of La Crosse and other arboviruses by *Aedes albopictus* and *Culex fatigans, Am. J. Trop. Med. Hyg.,* 24, 876, 1975.
193. **Watts, D. M., Pantuwatana, S., DeFoliart, G. R., Yuill, T. M., and Thompson, W. H.,** Transovarial transmission of La Crosse virus (California encephalitis Group) in the mosquito, *Aedes triseriatus, Science,* 182, 1140, 1973.
194. **Aitken, T. H. G., Tesh, R. B., Beaty, B. J., and Rosen, L.,** Transovarial transmission of yellow fever virus by mosquitoes (*Aedes aegypti*), *Am. J. Trop. Med. Hyg.,* 28, 119, 1979.
195. **Tesh, R. B., Chaniotis, B. N., and Johnson, K. M.,** Vesicular stomatitis virus (Indiana serotype): transovarial transmission by phlebotomine sandflies, *Science,* 175, 1477, 1972.
196. **Crane, G. T., Elbel, R. E., and Calisher, C. H.,** Transovarial transmission of California encephalitis virus in the mosquito *Aedes dorsalis* at Blue Lake, Utah, *Mosq. News,* 37, 479, 1977.
197. **Beck, M. D., Dodily, H., and O'Donnell, R.,** A strain of typhus rickettsiae isolated from the brain of a wild rat in California, *Public Health Rep.,* 59, 701, 1944.
198. **Lepine, P. and Sautter, V.,** Sur le duree de conservation du typhus murin dans l'encephale du spermophile, *Bull. Soc. Pathol. Exot.,* 29, 13, 1936.
199. **Philip, C. B. and Parker, R. R.,** The persistence of the viruses of endemic (murine) typhus, Rocky Mountain spotted fever and boutonneuse fever in tissues of experimental animals, *Public Health Rep.,* 53, 1246, 1938.
200. **Brigham, G. D.,** Endemic typhus virus in mice, *Public Health Rep.,* 53, 1251, 1938.
201. **Ignatovich, V. F.,** Experimental study of latency in typhus, in *Proc. 2nd Int. Symp. Rickettsiae Rickettsial Dis.,* Kazar, J., Ormsbee, R. A. and Tarasevich, J. N., Eds., Slovak Academy of Sciences, Bratislava, Czechoslovakia, 1978, 161.
202. **Audy, J. R., Harrison, J. L., and Wyatt-Smith, J.,** A survey of Jarak Island in the Malacca Straits, *Bull. Raffles Mus., Singapore,* 23, 230, 1950.
203. **Audy, J. R., Nadchatram, M., and Lim, B. L.,** Host distribution of Malayan ticks (Ixodoidea), Malaysian parasites, *Stud. Inst. Med. Res. Malaya,* No. 29, 225, 1960.
204. **Kodama, M., Kono, M., and Takahashi, K.,** Demonstration of *Rickettsia manchuriae* appearing in the stomach epithelial cells of rat fleas and rat lice infected with so-called Manchurian typhus, *Kitasato Arch. Exp. Med.,* 9, 91, 1932.
205. **Lepine, P.,** Recherches sur le typhus exanthematique et sur son origine murine, *Ann. Inst. Pasteur Paris,* 51, 290, 1933.
206. **Snyder, J. C. and Wheeler, C. M.,** The experimental infection of the human body louse, *Pediculus humanus corporis,* with murine and epidemic louse-borne typhus strains, *J. Exp. Med.,* 82, 1, 1945.
207. **Dyer, R. E.,** Effect of flea passage on epidemic typhus virus, *Public Health Rep.,* 49, 224, 1934.
208. **Fuller, H. S.,** Human body lice. IV. Direct serial passage of typhus rickettsiae by oral infection, *Proc. Soc. Exp. Biol. Med.,* 85, 151, 1954.
209. **Balaeva, N. M.,** The biological properties of *Rickettsia mooseri* after prolonged cultivation in the intestine of body lice. Communication IV. *J. Microbiol. Epidemiol. Immunobiol.* (*USSR*), 31, 2083, 1960.

210. **Price, W. H., Emerson, H., Nagel, H., Blumberg, R., and Talmadge, S.,** Ecologic studies on the interepidemic survival of louse-borne typhus fever, *Am. J. Hyg.*, 67, 154, 1958.
211. **Plotz, H.,** Complement fixation in rickettsial diseases, *Science*, 97, 20, 1943.
212. **Hamilton, H. L.,** Specificity of the toxic factors associated with the epidemic and the murine strains of typhus rickettsiae, *Am. J. Trop. Med.*, 25, 391, 1945.
213. **Craigie, J., Watson, D. W., Clark, E. M. and Malcomson, M. E.,** The serological relationships of the rickettsiae of epidemic and murine typhus, *Can. J. Res. Sect. E.*, 24, 84, 1946.
214. **Plotz, H., Bennett, B. L., Wertman, K., Snyder, M. J., and Gauld, R. L.,** The serological pattern in typhus fever. I. Epidemic, *Am. J. Hyg.*, 47, 150, 1948.
215. **Schwarz, E. and Schwarz, H. K.,** A monograph of the *Rattus rattus* group, *An. Esc. Nac. Cienc. Biol. Mexico City*, 4, 79, 1965 (1967).
216. **Pincus, G.,** A comparative study of the chromosomes of the Norway rat (*Rattus norvegicus* Erxleben) and the black rat (*Rattus rattus*), *J. Morphol.*, 54, 515, 1927.
217. **Schwarz, E.,** Classification, origin and distribution of commensal rats, *Bull. W.H.O.*, 23, 411, 1960.
218. **Pictet, F. J.,** Memoire sur des ossements trouves dans les graviers stratifies des environs de Mattegnin (Canton de Geneve), *Mem. Soc. Phys. Hist. Nat.*, 11, 85, 1846.
219. **Cornalia, E.,** Mammifers fossiles de Lombardie, in *Paleontologie Lombarde ou Description des Fossiles de Lombardie*, Ser. II, Stoppani, A., Ed., Antonio Stoppani, Milan, 1858, 1.
220. **Woldrich, J. N.,** Uber die diluviale Fauna von Zuzlawitz Winterberg im Bohmerwalde, *Sitzungsber. Akad. Wiss. Wien Math. Naturwiss. Kl. Abt. 1*, 84, 177, 1881.
221. **Woldrich, J. N.,** Diluviale Fauna von Zuzlawitz bei Winterberg im Bohmerwalde, *Sitzungsber. Akad. Wiss. Wien. Math. Naturwiss. Kl. Abt. 1*, 88, 978, 1883.
222. **Bate, D. M. A.,** New Pleistocene Murinae from Crete, *Annu. Mag. Nat. Hist. 11th Ser.* (*J. Nat. Hist.*), 9, 41, 1942.
223. **Bate, D. M. A.,** Pleistocene Murinae from Palestine, *Annu. Mag. Nat. Hist. 11th Ser.* (*J. Nat. Hist.*), 9, 465, 1942.
224. **Ellerman, J. F.,** The families and genera of living rodents. II. Family Muridae, *Br. Mus. Nat. Hist. Publ.*, 1, 1941.
225. **Misonne, X.,** African and Indo-Australian Muridae. Evolutionary trends, *Ann. Mus. l'Afr. Cent. Tervuren Belge. Ser. 8° Sci. Zool.*, 172, 1, 1969.
226. **Tchernov, E.,** *Succession of Rodent Faunas During the Upper Pleistocene of Israel, Mammalia Depticta*, Verlag Paul Parey, Hamburg, 1968, 1.
227. **Haas, G.,** Remarques sur la microfaune de mammiferes de la grotte d'Oumm Quatafa. Le Paleolithique et le Mesolithique du Desert de Judee, *Arch. Inst. Paleont. Hum. Mem.*, 24, 218, 1951.
228. **Haas, G.,** The fauna of layer B of the Abu Usba cave. The Abu Usba cave, Israel, *Explor. J.*, 2, 35, 1952.
229. **Davis, D. H. S.,** Subfossil house rats (*Rattus rattus*) from the iron age sites of Northern Rhodesia, *Bull. Zool. Soc. S. Afr.*, 3, 13, 1962.
230. **White, J. P. and O'Connell, J. F.,** Australian prehistory: new aspects of antiquity, *Science*, 203, 21, 1979.
231. **Sambon, L. W.,** Tropical and subtropical diseases, *J. Trop. Med. Hyg.*, 25, 170, 1922.
232. **Boelter, W. R.,** *The Rat Problem*, Bale and Danielsson, London, 1909, 1.
233. **Hinton, M. A. C.,** *Rats and Mice as Enemies of Man*, 3rd ed., British Museum of Natural History, 1931, 1.
234. **Koller, R.,** *Das Rattenbuch*, Schaper, Hanover, 1932, 1.
235. **MacArthur, W. P.,** The occurrence of the rat in early Europe. The plague of the Philistines (1 Sam., 5, 6), *Trans. R. Soc. Trop. Med. Hyg.*, 46, 209, 1952.
236. **Hirst, L. F.,** *The Conquest of Plague*, Clarendon Press, Oxford, 1953, 1.
237. **Toynbee, J. M. C.,** *Animals in Roman Life and Art*, Cornell University Press, Ithaca, N.Y., 1973, 1.
238. **MacArthur, W. P.,** Plague of the Philistines, *Trans. R. Soc. Trop. Med. Hyg.*, 46, 464, 1952.
239. **Rackham, J.,** *Rattus rattus:* the introduction of the black rat into Britain, *Antiquity*, 53, 112, 1979.
240. **MacArthur, W. P.,** The identification of some pestilences recorded in the Irish Annals, *Irish Hist. Stud.*, 6, 169, 1949.
241. **Gibbon, E.,** *The Decline and Fall of the Roman Empire*, Vol. 2, Smeaton, O., Ed., Modern Library, 1962, 1.
242. **Simpson, W. J.,** *A Treatise on Plague*, Cambridge University Press, England, 1905, 1.
243. **Shrewsbury, J. F. D.,** The plague of the Philistines, *J. Hyg.*, 47, 244, 1949.
244. **Mullo Weir, C. J.,** The background of the bible, in Manson, T. W., *A Companion to the Bible*, 2nd ed., Rowleg, H. H., Ed., T. and T. Clark, Edinburgh, 1963, 205.
245. **Osterley, W. O. E. and Robinson, T. H.,** *A History of Israel*, Clarendon Press, Oxford, 1932, 1.

246. **Zinsser, H.,** *Rats, Lice and History,* Little, Brown, Boston, 1935, 1.
247. **Hirsch, A.,** *Handbook of Geographical and Historical Pathology,* Vol. 1, (trans. Creighton, C), The New Sydenham Society, London, 1883, 1.
248. **MacArthur, W.,** The medical identification of some pestilences of the past, *Trans. R. Soc. Trop. Med. Hyg.,* 53, 423, 1959.
249. **Thucydides,** *The Peloponesian War,* Random House, New York, 1951, 110.
250. **Patrick, A.,** Disease in antiquity: Ancient Greece and Rome, in *Diseases in Antiquity,* Brothwell, D. and Sandison, A. T., Eds., Charles C Thomas, Springfield, Ill., 1967, 238.
251. **Crawfurd, R.,** *Plague and Pestilence in Literature and Art,* Clarendon Press, Oxford, 1914, 1.
252. **Marchette, N. J., Lundgren, D. L., Nicholes, P. S., Bushman, J. B., and Vest, D.,** Studies on infectious diseases in wild animals in Utah. II. Susceptibility of wild mammals to experimental plague, *Zoonoses Res.,* 1, 225, 1962.
253. **Blanc, G. and Baltazard, M.,** Recherches sur le mode de transmission naturelle de la peste bubonique et septicemique, *Arch. Inst. Pasteur, Maroc.,* 3, 173, 1945.
254. **Strong, R. P. and Teague, O.,** Studies on pneumonic plague and plague immunization. VII. Susceptibility of animals to pneumonic plague, *Philipp. J. Sci.,* 7 (Sect. B), 223, 1912.
255. **Shrewsbury, J. F. D.,** The plague of Athens, *Bull. Hist. Med.,* 24, 1, 1950.
256. **Page, D. L.,** Thycydides' description of the Great Plague at Athens, *Classical Quart.,* 47, 97, 1953.
257. **Manson, T. W.,** *A Companion to the Bible,* 2nd ed., Rowley, H. H., Ed., T. and T. Clark, Edinburgh, 1963, 1.
258. **Waley, A.,** *The Secret History of the Mongols and Other Pieces,* Brown and Noble, New York, 1964, 1.
259. **Heinrich, D.,** Remarks on the mediaeval occurrence of the Norwegian rat (*Rattus norvegicus* Berkenhaut, 1769) in Schleswig-Holstein, *Zool. Anz. Jena,* 196, 273, 1976.
260. **Lantz, D. E.,** The brown rat in the United States, Biol. Survey Bull. No. 33, U.S. Department of Agriculture, 1909, 1.
261. **Lamb, H.,** *The March of the Barbarians,* Doubleday, New York, 1940, 1.
262. **Brill, N. E.,** An acute infectious disease of unknown origin. A clinical study based on 221 cases, *Am. J. Med. Sci.,* 139, 484, 1910.
263. **Zinsser, H. and Castaneda, M. R.,** On the isolation from a case of Brill's disease of a typhus strain resembling the European type, *N. Engl. J. Med.,* 209, 815, 1933.
264. **Bozeman, F. M., Williams, M. S., Stocks, N. I., Chadwick, D. P., Elisberg, B. L., Sonenshine, D. E., and Lauer, D. M.,** Ecologic studies on epidemic typhus infection in the eastern flying squirrel, in *2nd Int. Symp. Rickettsiae Rickettsial Dis.,* Kazar, J., Ormsbee, R. A., and Tarasevich, I. N., Eds., Slovak Academy of Sciences, Bratislava, Czechoslovakia, 1978, 493.
265. **Sonenshine, D. E., Bozeman, F. M., Williams, M. S., Masiello, S. A., Chadwick, D. P., Stocks, N. I., Lauer, D. M., and Elisberg, B. L.,** Epizootiology of epidemic typhus (*Rickettsia prowazekii*) in flying squirrels, *Am. J. Trop. Med. Hyg.,* 27, 339, 1978.
266. **Woodman, D. R., Weiss, E., Dasch, G. A., and Bozeman, F. M.,** Biological properties of *Rickettsia prowazekii* isolated from flying squirrels, *Infect. Immun.,* 16, 853, 1977.
267. **McDade, J. E., Shepard, C. C., Redus, M. A., Newhouse, V. E., and Smith, J. D.,** Evidence of *Rickettsia prowazekii* infections in the United States, *Am. J. Trop. Med. Hyg.,* 29, 277, 1980.
268. **Buxton, P. A.,** *The Louse,* Williams & Wilkins Co., Baltimore, 1940, 1.
269. **Ewing, H. E.,** The sucking lice of American monkeys, *J. Parasitol.,* 24, 13, 1938.
270. **Ewing, H. E.,** Lice from human mummies, *Science,* 60, 389, 1924.
271. **Blanc, G. and Woodward, T. E.,** The infection of *Pedicinus albidus* Rudow, the maggot's louse, on typhus carrying monkeys (*Macaccus sylvanus*), *Am. J. Trop. Med.,* 25, 33, 1945.
272. **Arkwright, J. A., Bacot, A., and Martin Duncan, F.,** The association of rickettsia with trench fever, *J. Hyg.,* 18, 76, 1919.
273. **Atkin, E. E. and Bacot, A.,** Experiments on the infectivity of typhus virus contained in lice (*Pediculus humanus* and *Pedicinus longiceps*), *Br. J. Exp. Pathol.,* 3, 196, 1922.
274. **Wilder, R. M.,** The problem of transmission of typhus fever, *J. Infect. Dis.,* 9, 9, 1911.
275. **Anderson, J. R. and Goldberger, J.,** Studies on immunity and means of transmission of typhus, in Collected Studies on Typhus, Hyg. Lab. Bull. No. 86, U.S. Public Health Service, 1912, 81.
276. **Hare, R.,** *Pomp and Pestilence,* Vista Gollancz, London, 1954, 1.
277. **Audy, J. R.,** Man-made maladies, In *The Air We Breathe: A Study of Man and His Environment,* Farber, S. M. and Wilson, R. H. L., Eds., Charles C Thomas, Springfield, Ill., 1961, 100.
278. **Wisseman, C. L., Jr., Boese, J. L., Waddell, A. D., and Silverman, D. J.,** Modification of antityphus antibodies on passage through the gut of the human body louse, with discussion of some epidemiologic and evolutionary implications, *Ann. N.Y. Acad Sci.,* 266, 6, 1975.

INDEX

A

Aboimisalis, 42
Aborphysalis, 42
Abrocomidae, 60
Acarina, 30
  divergence of Argasidae and Ixodidae, 33
  origin, 30, 32—33
Acomys
  cahirimus, 67
  calurinus, *R. akari* and, 97
  distribution in Oriental region, 66—67
  typhus agglutinating antibody in, 121
Actinochitinose acarines, 30
Aegyptianella, 15, 21
  pullorum, 21, 38
Africa
  Rickettsia prowazekii in, 119—120
  rodents in, 59—60
  tick families in, 44
Age of reptiles, 30
Agglutinating antibody, to typhus rickettsiae, 120—122
Alectorobius canestrinii, see *Ornithodoros canestrinii*
Alloceraea, 42
Allodermanyssus sanguineus, 96
Allophysalis, 42
Alticola, distribution in Oriental region, 66
Amblyomma, 18, 20, 32
  experimental infection with *R. prowazekii*, 123
  origin, 70
  phylogeny and distribution, 39, 44
  R. rickettsii and, 91
Amblyomma americanum
  geographical distribution, 86—88
  R. rickettsii and, 86—88
Amblyomma americanum strain of rickettsiae, 77, 100
Amblyomma brasiliensis
  geographical distribution, 86, 88
  R. rickettsii and, 86, 88
Amblyomma cajennense, 89
  geographical distribution, 86, 87
  R. rickettsii and, 86, 87
Amblyomma cohaerens
  geographical distribution, 79
  R. conorii and, 79
Amblyomma cooperi
  geographical distribution, 86
  R. rickettsii and, 86
Amblyomma gemma
  geographical distribution, 79
  R. conorii and, 79
Amblyomma goeomydae, 39
Amblyomma hebraeum, 35
  geographical distribution, 79
  R. conorii and, 79
Amblyomma helvolum, 39
Amblyomma javanense, 39
Amblyomma maculatum, *R. parkeri* and, 92
Amblyomma ovale
  geographical distribution, 86
  R. rickettsii and, 86
Amblyomma striatum
  geographical distribution, 86, 88
  R. rickettsii and, 86, 88
Amblyomma variegatum
  geographical distribution, 79
  R. conorii and, 79
  R. typhi and, 117
Anactinochitinose acarines, 30
Anaplasma, 15, 21
Anaplasmataceae, see also specific genus, 8, 13, 15, 21—22, 76
  Aegyptianella, 15, 21
  Anaplasma, 15, 21
  Eperythrozoon, 16
  Haemobartonella, 15
  Paranaplasma, 15, 21
Anomaluridae, 59, 62
Anomaluroidea, 59—62
Anoplura, 141
Anteliomys, distribution in Oriental region, 66
Antibody
  R. conorii, 124
  R. prowazekii, origin in domestic animals, 124—125
Antigenic characteristics
  spotted fever rickettsiae, 76
  typhus rickettsiae, 114
Antricola, 31, 38
Aplondontidae, 59, 62
Aplodontoidea, 59—62
Apodemus, distribution in Oriental region, 66
Apomys, distribution in Oriental region, 66
Aponomma, 32, 33
  phylogeny and distribution, 38—39
Arboviruses, 130
Argas, 31, 38
  arboreus, 33, 81
  echinops, 38
  persicas, 21
Argasidae, see also specific genus
  Antricola, 31, 38
  Argas, 31, 38
  classification, 31
  divergence from Ixodidae, 33
  experimental infection with *R. prowazekii*, 123
  Ornithodoros, 31, 38
  Otobius, 31
  phylogeny and distribution, 38
Arthropods
  rickettsia association, 3—4
  R. conorii and, 79
  rickettsial evolution and, 24
Artiodactyla, 70
Arvicanthis niloticus

R. akari and, 97
R. typhi and, 115
Ascoschongastia indica, *R. typhi* and, 117
Asia
rodents in, 59
tick families in, 44
Athens plague, 138—139
Australia
rodent fauna of, 63—65
tick families in, 44
Aztec Indians, typhus in, 127

B

Babesia
bigemina, vestor, 40
bovis, 43
transovarial transmission, 35
Babesiidae, 35
Bacillus proteus, see also *Proteus vulgaris*
X 19, 3
Bacterial classification, 2
Baltica, 54
Bandicota, 65
distribution in Oriental region, 66—67
Bartonella baciliformis, 8, 21
geographical distribution, 2
Bartonellaceae, see also specific genus, 13, 15, 21
Bartonella, 15, 21
Grahamella, 15, 21
Bathyergidae, 60
Bathyergoidea, 58, 60 61
Batomys, distribution in Oriental region, 66
Bdellonyssus bacoti, see *Ornithonyssus bacoti*
Bedbug, see *Cimex*
Biphasic meningoencephalitis of Russia, 43
Blattabacterium, 15
cuenote, 20-21
Body lice, see *Pediculus humanus*
Boophilus, 35, 38
annulatus, 35, 39, 40
australis, see *Boophilus, microplus*
decoloratus, 35, 39, 40
geographical distribution, 79
R. conorii and, 79
geigyi, 39, 40
kohlsi, 39
microplus, 35, 39, 40
R. typhi and, 115, 117
phylogeny and distribution, 39—40
Borrelia, 34
duttoni, 34
hispanica, 34
recurrentis, 34
Pediculus humanus and, 141
transovarial transmission, 35
Boutonneuse fever, see also *Fievre boutonneuse*,
96, 97, 119
Brachylagus, geographical distribution, 90
Brill's disease, 4
Brill-Zinsser disease, 125, 143
Bubonic plague, history, 136, 137
Bullis fever, 100

C

California encephalitis virus, 130
Cambrian period, 32
geological evolution during, 55
Caprolagus, geographical distribution, 90
Capromyidae, 59
Carboniferous period, 30, 32, 33
Carpomys, distribution in Oriental region, 66
Castoridae, 59, 62
Castoroidea, 59—62
Caviamorph rodents, 61
Caviidae, 59
Cavioidea, 58—61
Celaenomys, distribution in Oriental region, 66
Cell wall
chlamydial, 14
rickettsial, 14
R. prowazekii, 13, 14
R. rickettsii, 14
Cenozoic era, 30, 32, 33
geological evolution during, 55
Ceratomys, distribution in Oriental region, 66
Ceratophyllus
fasciatus, *R. typhi* and, 126
tesquorum, geographical distribution, 83
R. sibirica and, 83
Chiggerborne rickettsiosis, see also
Tsutsugamushi fever, 8
Chiggers, 8
China, 54
Chinchillidae, 59
Chinchilloidea, 58—61
Chiromyscus, distribution in Oriental region, 66
Chiropodomys, distribution in Oriental region,
66, 67
Chlamydia
psittaci, 22
trachomatis, 22
Chlamydiaceae, see also specific genus, 13, 22
Chlamydia, 22
Colesiota Rake, 22
Ricolesia Rake, 22
Chlamydiae
cell wall, 14
classification, 14
Bergey's Manual, 13
Chlamydiales, 12, 13, 22
Chlamydozoaceae, 12
Chrotomys, distribution in Oriental region, 66
Cimex, 20
Citellus lateralis, *R. rickettsii* and, 88
Classical typhus, see Epidemic typhus
Clenodactylidae, 60
Coelomys, distribution in Oriental region, 66
Colestiota Rake, 22
Commensalism, 23
Conquistadores, epidemic typhus and, 127
Continental drift, 54

Cowdria, 15, 18, 20
ruminantium, 18, 20, 35
Coxiella burnetii, 16, 35, 38, 43, 100
immunological characteristics,
Crecetinae, 63
Cretaceous period, 32
geological evolution during, 55
Cricetidae, 59, 63
Crunomys, distribution in Oriental region, 66
Ctenocephalus
canus, *R. typhi* and, 126
felis, endemic typhus and, 116
R. typhi and, 126
Ctenodactylae, in India, 65
Ctenodactyloidea, 58, 60, 61
Ctenomyide, 60
Cyanobacteria, 13
Cytoecetes ondiri, 18

D

Dacnomys, distribution in Oriental region, 66
Dasyproctidae, 59
Dephomys, 133
Dermacentor, 32, 35
andersoni, 20, 41
experimental infection with *R. prowazekii*, 123, 124
geographical distribution, 85, 86
hosts, 88
R. montana and, 92
R. rickettsii and, 81, 85, 86
Sylvilagus nuttalli and, 89
astrosignatus, 41
auratus, 41
geographical distribution, 79
R. conorii and, 79
compactus, 41
geographical distribution, 44
marginatus, 35
geographical distribution, 79, 82, 83, 86
R. conorii and, 79
R. rickettsii and, 86
R. sibirica and, 82—84
nitens, 35
experimental infection with *R. prowazekii*, 123
nitens Neumann, 86
geographical distribution, 86
R. rickettsii and, 86
nutalli, 82
geographical distribution, 82, 83
R. sibirica and, 82, 83
occidentalis, 86
geographical distribution, 86
R. rickettsii and, 86
Sylvilagus bachmanni and, 89
parumapertus, 41
geographical distribution, 86
hosts, 88
Lepus californicus and, 89
R. rickettsii and, 86
phylogeny and distribution, 41
pictus, 79
experimental infection with *R. prowazekii*, 123
geographical distribution, 79, 82, 83
R. conorii and, 79
R. sibirica and, 82, 83
reticulatus, 79
geographical distribution, 79
R. conorii and, 79
silvarum, 35
geographical distribution, 82, 83
R. sibirica and, 82, 83
variabilis, 41
geographical distribution, 86, 87
R. montana and, 92
R. rickettsii and, 86, 87
Sylvilagus floridanus and, 89
Dermacentor occidentalis strain of rickettsiae, 77, 99
Dermacentor parumapertus strain of rickettsiae, 77, 99
geographical distribution, 99
Dermacentor variabilis strain of rickettsiae, 77
Dermacentroxenus, 16, 76
evolution, 100—102
R. typhi and, 130
rickettsii var. *conori*, 78
sibiricus, 82
Dermaphysalis, 42
Devonian period, 30, 32
Didelphis marsupialis, *R. rickettsii* and, 88
Dinomyidae, 59
Dipodidae, 59
Dipodoidea, 59—61, 63
DNA hybridization studies, typhus complex, 114

E

Echidnophaga gallinacea, *R. typhi* and, 116
Echimyidae, 60
Echiothrix, distribution in Oriental region, 66
Egypt, *R. prowazekii* in, 121
Ehrlichia, 15, 18
bovis, 18
canis, 18, 23
equi, 18
geographical distribution, 18
ovina, 18
phagocytophia, 18
Ehrlichieae, see also specific genus, 7, 13, 15, 18, 20
Cowdria, 15, 18, 20
Ehrlichia, 15, 18
Neorickettsia, 15, 20
Emerods, see Inguinal buboes
Emys leprosa, 21
Endemic relapsing fever, 34
Endemic typhus, 4, 8, 114, 115
Endopalpiger, 43
Eocene epoch, 32, 33

geological evolution during, 55
Rodentia in, 57
Eothenomys, distribution in Oriental region, 66
Eperythrozoon, 16
Epidemic relapsing fever, 34
Epidemic typhus, 4, 8, 114, 118
interepidemic maintaining host, 125
man-made malady, 142—143
Erethizontidae, 58, 59
Erethizontoidea, 58—61
Eropeplus, distribution in Oriental region, 66
Ethiopia, *R. prowazekii* in, 120—121
Ethiopian region
Muroidea in, 63
Rhipicephalus sanguineus in, 81
sciurids in, 62
Ethnobotany, 128
Eurasia, rodents in, 59
Europe
history of plague in, 135
migration of rodents to, 133—135
rodents in, 59
European tickborne encephalitis, 43
Eutamias amoenus, *R. rickettsii* and, 88
Exopalpiger, 43

F

Faunal regions ofworld, 17
Fievre boutonneuse, see also Boutonneuse fever, 4, 7, 78, 80
Fiji epidemic, 138
Filariae, 130
Fleaborne typhus, see Endemic typhus; Murine typhus
Flying squirrels, 62, 65
R. prowazekii and, 118, 140—141
Fossils
hystricomorphs, 58, 61
muroids, 63
Rodentia, dstribution, 61
sciurids, 62
Franciscella tularensis, 100
geographical distribution, 2
Funambulus, in India, 65

G

Gamasoid mites, see Mesostigmata
Geological evolution, 54—56
breakup of Pangaea, 55—56
plate tectonics and seafloor spreading, 54—55
Geological time divisions, 32
Geomyoidea, 59—62
Gerbillus, distribution in Oriental region, 66, 67
Glaucomys, 62
volas volans, see Flying squirrels
Gliridae, 59
Gliroidea, 59—61, 63
Golunda
distribution in Oriental region, 66
ellioti, *R. typhi* and, 115
Gondwana, 54
Gondwana system, 67
Gondwanaland supercontinent, 55, 67
Amblyomma in, 39
Goxixodes, 42
Grahamella, 21
Ground squirrels, 65
Gut-wall barrier, 129

H

Hadromys, distribution in Orietal region, 66
Haemaphysalis, 35
bispinosa, 35
geographical distribution, 79
R. conorii and, 79
concinna, 82
geographical distribution, 82, 83
R. sibirica and, 82, 83
geographical distribution, 44
humerosa, 35
japonica, 82
geographical distribution, 82, 83
R. sibirica and, 82, 83
leachi, 35, 41
geographical distribution, 79, 80, 86
R. conorii and, 79, 80, 78
R. rickettsii and, 86
leporispalustris, 5
geographical distribution, 86, 88
host, 88
R. canada and, 116, 117
R. rickettsii and, 86, 88
papuana, 79
geographical distribution, 79
R conorii and, 79
phylogeny and distribution, 41—42
punctata, 83
geographical distribution, 83
R. sibirica and, 83
subgenera, 42
theilerae, 41
Haemaphysalis leporispalustris strain of rickettsiae, 77
Haemobartonella, 15
Haemophysalis, 42
Haeromys, distribution in Oriental region, 66
Heartwater fever, 18, 35
Hepalomys, distribution in Oriental region, 66
Herpes simplex virus, 130
Herpes varicella-zoster virus, 130
Herpetobia, 42
Heteromyidae, 59
Hlp-like *R. rickettsii*, 99
Host-parasite dynamics, 94
Hyalomma, 35
aegyptium, 79
geographical distribution, 79
R. conorii and, 79

albiparmatum, 79
geographical distribution, 79
R. conorii and, 79
anatolicum, experimental infection with *R. prowazekii*, 123
detritum, 82
geographical distribution, 82, 83
R. sibirica and, 82, 83
dromedarii, 35
experimental infection with *R. prowazekii*, 124
geographical distribution, 83
R. canada, and, 117
R. sibirica and, 83
excavatum, 35
geographical distribution, 83
R. sibirica and, 83
geographical distribution, 44
marginatum, 35
gographical distribution, 79, 82, 83
R. conorii and, 79
R. sibirica and, 82, 83
phylogeny and distribution, 40
R. typhi and, 115
Hybomys, 33
Hydrochoeridae, 58, 59
Hydromyinae, 63, 64
Hystricidae, 59
in India, 65
Hystricoidea, 58—61
Hystricomorpha, 58, 60—61

I

India
R. conorii in, 80
rodent fauna of, 65—67
Indian tick bite fever, 4
Indian tick typhus, 40
Indocentor, 38
Indo-Malayan region
rodents in, 59
sciurids in, 62
Inguinal buboes, 136
Insectborne rickettsiae, 4—5, 8, 18
Ixodes, 35
apronaphorus, 83
geographical distribution, 83
R. sibirica and, 8
brunneus, 86
geographical distribution, 86
R. rickettsii and, 86
cookei, 86
geographical distribution, 86
R. rickettsii and, 86
dentatus, 86
geographical distribution, 86
R. rickettsii and, 86
geographical distribution, 44
granulatis, 43
granulatus, 79
geographical distribution, 79
R. conorii and, 79
hexagonus, 43
geographical distribution, 79
R. conorii and, 79
holocyclus, 43
geographical distribution, 98
Queensland tick typhus and, 98
loricatus, 86
geographical distribution, 86
R. rickettsii and, 86
lunatus, 41
origin, 70
persulcatus, 35, 43
geographical distribution, 83
R sibirica and, 83
phylogeny and distribution, 42—43
plumbeus, 83
geographical distribution, 83
R. sibirica and, 83
ricinus, 18, 32, 35, 43
geographical distribution, 79, 80, 83
R. conorii, and, 79, 80, 78
R. sibirica and, 83
scapularis, 86
geographical distribution, 86
R. rickettsii and, 86
tasmani, geographical distribution, 98
texanus, 86
geographical distribution, 86
R. rickettsii and, 86
Ixodes pacificus strain of rickettsiae, geographical distribution, 77, 99
Ixodidae, see also specific genus, 31, 37, 38—43
Amblyomma, 31, 39
Anomalohimalaya, 31
Aonomma, 31, 38—39
Boophilus, 31, 39—40
Cosmiomma, 31, 38
Dermacentor, 31, 41
divergence from Argasidae, 33
experimental infection with *R. prowazekii*, 123
Haemaphysalis, 31, 41—42
Hyalomma, 31, 38, 40
Ixodes, 31, 42—43
Margaropus, 39
Nosomma, 31, 38
Rhipicentor, 31, 38
Rhipicephalus, 31, 40
Ixodoidea, classification, 31

J

Japanese encephalitis virus, 130
JC 880, 77
Jurassic period, 32,33
geological evolution during, 55
Justinian plague, 135

K

Kaiseriana, 42

Karroo system, 67
Kazakhstania, 54
Kenyan tick bite fever, 4
Kyasanur Forest disease virus, 42

L

LaCrosse virus, 130
Lagomorpha, 90
Lagomorphs
  origin, 90
  role in *R. rickettsii* ecology and evolution, 88—91
Land bridges, 57, 61, 90, 91, 101, 140
Langat virus, 43
Laurasia, 55
Laurentia, 54
Leiognathus bacoti, see *Ornithonyssus bacoti*
Leishmania, 130
Lenomys, distribution in Oriental region, 66
Leporidae, see also specific genus
  Brachylagus, 90
  Caprolagus, 90
  geographical distribution, 90
  Lepus, 89, 90
  Nesolagus, 90
  origin and evolution of *R. rickettsii*, 89—91
  Oryctolagus, 90
  Pentalagus, 90
  Pronolagus, 90
  Romerolagus, 90
  spotted fever in, 88
  Sylvilagus, 87—90
  Tapeti, 90
Leptopsylla musculi, *R. typhi* and, 115, 126
Lepus
  americanus, *R. canada* and, 116
    R. rickettsii and, 88
  californicus, *R. canada* and, 116
    R. rickettsii in, 88
  cuniculus, see *Cryctolagus cuniculus*
  geographical distribution, 90
  timidus, 89
L-forms, 14
Limnomys, distribution in Oriental region, 66
Liponyssoides sanguineus, 30, 96
Liponyssus bacoti, see *Ornithonyssus bacoti*
Litomosoides carinii, 128
Lone Star virus, 100
Lophuromys flavopunctatus, *R. typhi* and, 115
Louping ill, 43
Louseborne typhus, see also Epidemic typhus, 4, 8
  tickborne member, 5

M

Macronyssus bacoti, see *Ornithonyssus bacoti*
Maculatum agent, see *Rickettsia, parkeri*
Maintaining host, 95
Malignant canine rickettsiosis, 18
Malomys, distribution in Oriental region, 66
Margaropus, phylogeny and distribution, 39
Mastomys, 133
  natalensis, *R. typhi* and, 115
Matlalzahuatl, see Tabardillo
Measles, 138
Melasmothrix, distribution in Oriental region, 66
Melomys, 64
Meriones, distribution in Oriental region, 66
Mesostigmata, 7, 30
  origin, 30
Mesozoic era, 30
  geological evolution during, 54
Metastigmata, 7, 30
  classification, 31
  origin, 30
  phylogeny and distribution, 37—43
Mexican typhus, see Tabardillo
Microagglutination test, *R. prowazekii*, 120
Microbial symbiotes, of ticks, 33—36
Micromys, distribution in Oriental region, 66
Microtatobiotes, 12, 13
Microtus
  distribution in Oriental region, 66
  pennsylvanicus, *R. rickettsii*, and, 88
  R. montana and, 92
Middle East
  history of plague in, 135—137
  migration of rats to, 133—134
Milardia kondara, *R. typhi* and, 115
Millardia, distribution in Oriental region, 66
Miocene epoch, 32
  geological evolution during, 55
Mites, see also Acarina, *Ornithonyssus*
  as original host for *R. typhi*, 128—129
Molecular biology, typhus rickettsiae, 114
Mollicutes, 13
Muramic acid, 14
Muridae, 57, 59, 63—67
  genera distribution in Oriental region, 66
Murinae, 63
Murine typhus, 8, 78, 114, 115, 126
Muroidea, 59, 60, 61, 63—67
Mus, 63, 65
  booduga, 67
    R. typhi and, 115
  distribution in Oriental region, 66—67
  leggadilla, 67
  musculus, *R. akari* and, 96
    R. typhi and, 115
  platythrix, *R. typhi* and, 115
  saxicola, *R. typhi* and, 115
  wagneri, *R. typhi* and, 115
Mutualism, 23
Mycoplasmas, 12, 13
Mycoplasmatales, 12, 13
Mycteromys, distribution in Oriental region, 66
Myomorpha, 59—61, 63—67
  Australian rodent fauna, 63—65
  Oriental region rodent fauna, 65—67
    India, 65, 67
    Philippines, 67

Southeast Asia, 67
Myomys, 133

N

Natural selection, 56
Nearctic region
Muroidea in, 63
R. rickettsii in, 85
Rodentia in, 58
sciurids in, 62
Neodon, distribution in Oriental region, 66
Neohaematopinus sciuropteri, *R. prowazekii* and, 140
Neorickettsia, 15, 20
helminthoeca, 20, 22, 23
sennetsu, 22
Neotropical region
R. rickettsii in, 85
Rodentia in, 58
evolution, 57
sciurids in, 62
Nesokia
distribution in Oriental region, 66
indica, 82
Nesolagus, geographical distribution, 90
Neurotropic viruses, 130
Nonphototropic bacteria, see Scotobacteria
North America
rodents in, 59
tick families in, 44
North Asian (Siberian) tick typhus, 7, 82

O

Ochromys, 133
Octodontidae, 60
Octodontoidea, 58, 60, 61
Oligocene epoch, 32
Onychopalpida, 30
Opaliid ciliate infusoria, 25
Orchopeas howardii, *R. prowazekii* and, 140
Ordovician period, 32
Oriental rat flea, see *Xenopsylla cheopis*
Oriental region
rodent fauna of, 65—67
sciurids in, 62
tick families in, 44
Ornithodoros, 31, 38
canestrinii, experimental infection with *R. prowazekii*, 123
moubata, 34, 35
experimental infection with *R. prowazekii*, 123
R. canada and, 117
nicollei, 86
geographical distribution, 86
R. rickettsii and, 86
papillipes, *R. canada* and, 117
parkeri, 86
geographical distribution, 86
R. rickettsii and, 86
rostralus, 86
geographical distribution, 86
R. rickettsii and, 86
rudis, 86
geographical distribution, 86
R. rickettsii and, 86
savignyi, 35
Ornithonyssus
bacoti, 30
R. akari and, 97
R. typhi and, 115, 116, 128—129
ondatrae, 129
Ornithophysalis, 42
Oroya fever, 21
Oryctolagus
cuniculus, 40, 89
geographical distribution, 90
Otobius lagophilus
geographical distribution, 86
R. rickettsii, 86
Otomys, *R. typhi* and, 115

P

Palearctic region
Myomorpha in, 63
Rhipicephalus sanguineous in, 81
Rodentia in, 58
evolution, 56
sciurids in, 62
Paleocene epoch, 32
geological evolution during, 55
Rodentia in, 57
Paleozoic era, 30
geological evolution, 54
Pangaea supercontinent, 54
breakup of, 55
Papuan-Australian rodent fauna, 64
Paramyidae, 57
Paranaplasma, 15, 21
Parapediculus, 141
Parasitiformes, 30
Parasitoid mites, see Mesostigmata
Pasteurella pestis, see *Yersinia pestis*
Pedetidae, 59, 70
Pedicinus, 141
albidus, 141
longiceps, 141
Pediculidae, see also specific genus
Pedicinus, 141
Pediculus, 141
Phthirus, 141
Pediculus humanus, 114
epidemic typhus and, 118
R. prowazekii and, 141—142
var. *capitis*, 118, 141
var. *corporis*, 118
Pentalagus, geographical distribution, 90

Permian period, 32, 33
  geological evolution during, 54, 55
Peromyscus
  leucopus, R. montana and, 92
    R. prowazekii and, 140
    R. rickettsii and, 88
  polionotus, R. typhi and, 115
  R. montana and, 92
Petromyidae, 60
Phaiomys, distribution in Oriental region, 66
Philippines, rodent fauna, 67
Philistine plague, 136, 137
Phlebotomus fever virus group, 130
Phloeomys, distribution in Oriental region, 66
Photobacteria, 13
Phthirus, 141
Phylogenetic relationships, rickettsiae, 6—8
Pithecheir, distribution in Oriental region, 66
Pitymys pinetorum
  R. montana and, 92
  R. rickettsii and, 88
Plague
  Athens, 138—139
  bacilli, 130
  history, 135—137
Platacanthomyidae, 59
Platacanthomys lasiurus, in India, 65
Plate tectonics, 54
Pleistocene epoch, 32
Pliocene epoch, 32
  geological evolution during, 55
Pneumonic plague, 138
Polyplax spinulosus
  R. prowazekii and, 131
  R. typhi and, 114, 115
Popillia japonica, 20
Praomys, 133
Primary host, 95
Procaryotae, 13
Pronolagus, geographical distribution, 90
Proteus, 3
  OX 19, 4
  OX K, 4
  vulgaris, 3
    Kingsbury strain, 3—4
  X, 126
Proteus nonagglutinating organisms, 6
Proteus OX agglutination test, 4
Proteus OX-19 agglutinating antibodies, typhus rickettsiae and, 118, 120
Protophyta, 13
Pseudomyinae, 63, 64
Pseudomys, 64
Pulex irritans, R. typhi and, 116
Pyromys, distribution in Oriental region, 66

Q

Quarternary period, 32
Queensland tick typhus, 7, 98

R

Rabies virus, 130
Rat, see *Rattus*
Rattus, 64, 65
  blandforde, R. typhi and, 115
  distribution in Oriental region, 66, 67
  edwardsi, 133
  exulans, geographical distribution, 132
    R. typhi and, 115
  kiridus, 133
  norvegicus, 133
    geographical distribution, 132
    origin and distribution, 139—140
    R. akari and, 97
    R. prowazekii and, 122, 131
  rattus
    alexandrinus, 132, 134
    argentiventer, 132, 139
    caraco, 132
    diardi, 132, 133
    exulans, 132
    kandianus, 132
    nitidus, 132
    origin and dispersion, 132—137
    rattus, 132
    R. akari and, 97
    R. typhi and, 115, 128, 130—131
    roquei, 132
    rufescens, 132
    satarae, R. typhi and, 115
    wichmanni, 132
  typhus agglutinating antibodies in, 122
Ratufa, in India, 65
Recrudescent typhus, 125, 140, 143
Relapsing fevers, 34, 141
Reservoir, 94—95
Rhipicephalus, 18, 35, 38
  appendiculatus, 40
    geographical distribution, 79
    R. conoii and, 78, 79
  bursa, 35
  evertsi, 40
    geographical distribution, 79
    R. conorii, geographical distribution, 44
  guilhoni, 40
  haemaphysaloides, geographical distribution, 79
    R. conorii and, 79
  origin, 70
  phylogeny and distribution, 40
  pulchellus, geographical distribution, 79
    R. conorii and, 79
  pusillus, 40
  sanguineus, 18, 20, 35, 40, 95
    geographical distribution, 79, 80, 82, 83, 86
    R. conorii and, 78—80
    R. sibirica and, 82, 83
    R. rickettsii and, 86
  schulzei, geographical distribution, 83
    R. sibirica and, 83
  simus, 40
    geographical distribution, 79
    R. conorii and, 79
    R. typhi and, 117
  sulcatus, 40

turanicus, 40
geographical distribution, 83
R. sibirica and, 83
Rhipistoma, 42
Rhizomyidae, 59, 63, 65
Rhynchomys, distribution in Oriental region, 66
Rickettsemia, in domestic animals, 122—123
Rickettsia
akari, 7, 96
affinities, 97
evolution, 101, 102
geographical distribution, 19, 77, 96
immunological group, 5
natural hosts, 96—97
reptiles and, 38
rickettsialpox and, 96
subgrouping, 6
vector, 30
australis, 7, 35, 97
affinities, 98
evolution, 101—102
geographical distribution, 16, 19, 77
immunological group, 5
natural hosts, 98
origin, 98
subgrouping, 6
vector, 43
canada, 7
antigenic characteristics, 114
geographical distribution, 16, 19, 77, 116
host range, 116—118
immunological characteristics, 5
immunological group, 5
molecular biology, 114
R. typhi and, 130
conorii, 7
agglutinating antibody to, 120
antibody, 120, 124
arthropod hosts, 79
center of origin, dispersion from, 81—82
classification according to *Proteus* OX agglutination, 4
evolution, 100—101
geographical distribution, 16, 19, 77, 78
hereditary transmission, 81
immunological group, 5
in India, 80
natural cycles in Africa, 78
reptiles and, 38
subgrouping, 6
vectors, 39, 40, 42
manchuriae, 131
montana, 7
geographical distribution, 77, 92
hosts, 92—93
immunological group, 5
subgrouping, 6
parkeri, 7
evolution, 101
geographical distribution, 16, 77, 92
host, 92
immunological group, 5
subgrouping, 6
prowazekii, 8, 35, 62, 142—143
agglutinating antibody to, 120
antibody, origin in domestic animals, 124—125
antigenic characteristics, 114
cell wall, 13
classification according to *Proteus* OX agglutination, 4
derivation from *R. typhi*, 131
evolution, 131—143
experimental infection of domestic animals, 122—123
experimental infection of ticks, 123—124
flying squirrels and, 140—141
geographical distribution, 18, 118
history of early typhus epidemics, 137—139
host range, 118—125
human body louse, 141—142
immunological characteristics, 4—5
immunological group, 5
interepidemic maintenance in man, 125
molecular biology, 114
Pediculus humanus and, 141—142
presence in ticks and domestic animals, 119—121
Rattus norvegicus and, 139—140
relationship to *R. typhi*, 126—127
reptiles and, 38
R. canada vs., 5
separation from *R. typhi*, 118—119
typhus and, 137, 142—143
var. *mooseri*, 78
rhipicephali, 7, 95—96
geographical distribution, 77
immunological group, 5
rickettsii, 7
arthropod hosts and their distribution, 86
acarine hosts, 85—88
cell wall, 14
classification according to *Proteus* OX agglutination, 4
Dermacentor andersoni and, 81
ecology and evolution, role of lagomorphs in, 88—91
evolution, 101
geographical distribution, 16, 19, 77, 85
Hlp-like strain, 99
host distribution, immunological group, 5
infections, domestic animals and, 89
natural cycle, 89
origin and dispersion, 91
subgrouping, 6
transovarial transmission, 81
vectors, 41, 42, 87, 91
wild vertebrate hosts, 88—91
sibirica, 7
arthropod hosts and their distribution, 83, 84
evolution, 100—101
geographical distribution, 16, 19, 77, 82, 84, 85
immunological group, 5
subgrouping, 6
vectors, 41, 42, 84

slovaca, 7, 83, 84
geographical distribution, 19, 77
immunological group, 5
subgenera, 16
tsutsugamushi, 8
antigenic variants, 5
classification according to *Proteus* OX agglutination, 4
geographical distribution, 2, 18
immunological group, 5
typhi, 8, 142
agglutinating antibody to, 120
antigenic characteristics, 114
classification according to *Proteus* OX agglutination, 4
derivation of *R. prowazekii* from, 131
evolution, 127—131
geographical distribution, 18, 114
host range, 114—116
immunological characteristics, 4—5
immunological group, 5
molecular biology, 114
origin, 127—128, 144
rats and, 130—131
relationship to *R. prowazekii*, 126—127
reptiles and, 38
R. canada and, 130
R. canada vs., 5
separation from *R. prowazekii*, 118—119
Xenopsylla cheopis and, 129—130
Rickettsiaceae, 12—21
definition, 22—23
tribe Ehrlichieae, 18—20
Cowdria, 18, 20
Ehrlichia, 18
Neorickettsia, 20
tribe Rickettsieae, 16, 18
Coxiella, 15
insectborne, 18
Rickettsia, 15
Rochalimaea, 15
tickborne, 16
trombiculid miteborne, 18
tribe Wolbachieae, 15, 20—21, 76
Blattabacterium, 15, 20—21
Rickettsiella, 15, 20—21
Symbiotes, 15, 20—21
Wolbachia, 15, 20
Rickettsiae
Amblyomma americanum strain, 77
cell wall, 14
classification, 3—6, 15
according to vector, 6—8
arthropod association and pathogenicity, 3—4
Bergey's Manual, 13
evolution, 3—6
immunological characteristics, 4—6
natural, 8
phylogenetic relationships, 6
present system, 12—14
Proteus nonagglutinating organisms, 6
Dermacentor occidentalis strain, 77, 99
Dermacentor parumapertus strain, 77, 99
Dermacentor variabilis strain, 77
differentiation from other bacteria, 12—14
distribution and relationships, 11—25
geographical distribution, 16, 18—19
Haemaphysalis leoprispalustris strain, 77
insectborne, 4—5, 8, 18
Ixodes pacificus strain, 77
origin and evolution, 23—25, 36—37
relationship with ticks, evolution, 29—45
separation from viruses, 12
spotted fever group, 75—102
tickborne, 5—7, 16, 37
trombiculid miteborne, 5, 8, 18, 37
typhus complex, see also Typhus rickettsiae; specific rickettsiae, 114—144
Rickettsiales, 12, 13
Rickettsialpox, 7, 96
Rickettsieae, see also specific genus, 13, 15, 16, 18
Coxiella, 15
Rickettsia, 15, 16, 18
Rochalimaea, 15
Rickettsiella, 15, 20—21
popilliae, 20
Rickettsiosis, 7
malignant canine, 18
Rickettsoideae, 12
Ricolesia Rake, 22
Rochalimaea quintana, 3, 8, 15, 16, 23
immunological characteristics, 6
Rocky Mountain spotted fever, 3, 4, 7, 85, 87
rabbit distribution and, 89
R. canada and, 116
vector, 40
Rodents, see Rodentia
Rodentia
classification and distribution, 58—68
Hystricomorpha, 58, 60—61
Myomorpha, 60, 61, 63—67
Sciuromorpha, 60, 61—63
evolution, 54, 56
center of origin, 56—57
time of origin, 57—58
geographical distribution, 69
living families and probable regions of origin, 59—60
R. typhi in, 116
superfamilies, geographical distribution, 60, 61
Romerolagus, geographical distribution, 90
Rural (chiggerborne) typhus, 115
Russian spring summer encephalitis, 41, 43

S

Sao Paulo fever, 4
Sarcoptiform mites, 30
Sarna en el rebono, see Tabardillo
Schizomycetes, 13
Schizophyceae, 13
Sciuridae, 57, 59, 62

in India, 65
Sciuroidea, 59—62
Sciuromorpha, 59—63
Scotobacteria, 12, 13
Scrapie virus, 130
Scrub typhus, 4
Seafloor spreading, 54
Seleviniida, 59
Sharifiella, 42
"Shop" typhus, see Endemic typhus
Siberia, 54
Siberian tick typhus, 82
Sigmodon hispidus, 128
R. rickettsii and, 88
Silurian period, 32
Smallpox, 138
South African fleaborne (endemic) typhus, 78
South African tick-bite fever, 4, 78
South African tick typhus, 78
South America
rodents in, 59—60
tick families in, 44
Southeast Asia
rodent fauna in, 67
tick typhus rickettsiae in, 80
Spalacidae, 59, 63
Spirochaetes, 130
vectors of, 34
Spotted fever, 80
rickettsiae, see also specific rickettsia, 75—102
antigenic characteristics, 76
classification, 77
evolution of *Dermocentroxenus* line, 100
geographical and host distribution, 77, 78
Sternalixodes, 43
Stochomys, 133
Suncus
murinus, *R. typhi* and, 115
typhus agglutinating antibody in, 122
Swiss agent, 77, 78
Sylvilagus, 88
audubonii, 88
bachmanni, 89
floridanus, 88, 89
geographical distribution, 90
minensis, 87
nuttalli, 89
R. rickettsii and, 88—90
Symbiosis, 23
Symbiote, 23
microbial, of ticks, 33—36
Symbiotes, 15, 20—21
lectularis, 20

T

Tabardillo, 4, 126—127
Tache noire, 95, 96
Tapeti, geographical distribution, 90
Tarsomys, distribution in Oriental region, 66
Tatera
distribution in Oriental region, 66, 67
robusta, *R. typhi* and, 115
Taxonomy
bacteria, 2
chlamydiae, 22
rickettsia, evolution, 3—6
Rickettsiales, see also specific family, 14—22
family Anaplasmataceae, 15, 21—22
family Bartonellaceae, 15, 21
family Rickettsiaceae, 14—21
Terminology, 6—7
Tertiary period, 32
Tethys Sea, 55
Texas fever, 40
Thai TT—118, 77
Theileriidae, 35
30 years war, typhus during, 118
Thryonomyidae, 60
Tick(s), see also Argasidae; Ixodidae; Metastigmata, 30
geographical distribution, 44
hard-bodied, see Ixodidae
microbial symbiotes of, 33—36
phylogeny and distribution, 37—43
soft-bodied, see Argasidae
Tickborne fever of cattle and sheep, 43
Tickborne paroxysmal rickettsiosis of Ukrainia, 43
Tickborne rickettsiae, 5—8, 16, 37
classification, 77—78
evolution, 100—102
spotted fever and tick typhus group, 75—102
Tickborne typhus, 5—6
Tick-rickettsia relationship
evolution of, 29—45
origin and development, 36—37
Tick typhus rickettsiae, in Southeast Asia, 80
Tlalzhuatl, see *Trombicula irritans*
Trench fever, 8
Triassic period, 32, 33
geological evolution during, 54, 55
Trombicula irritans infestation, 127
Trombiculid miteborne rickettsiae, 5, 8, 18
Trombiculid miteborne typhus, 5
Trombidiform mites, 30
Tryphomys, distribution in Oriental region, 66
Tsutsugamushi fever, 3, 4
TT-118 strain, 82
Typhlomys, 65
Typhus, 3
antibodies, origin in domestic animals, 124—125
classification according to *Proteus* OX agglutination, 4
epidemics, history of, 137—139
fever, vectors, 3
historical notes, 126—127
rickettsiae, 113—143, see also specific rickettsiae
characteristics, 114
evolution of *R. prowazekii*, 131—143
evolution of *R. typhi*, 127—131

geographical distribution and host range, 114—125
origin, 126—143
relationship of *R. prowazekii* to *R. typhi*, 126

U

Urban typhus, see Endemic typhus
Uromys, 64

V

Vandeleuria, 65
distribution in Oriental region, 66
Vectors
Aegyptianella pullorum, 21
Babesia bigemina, 40
central European tickborne encephalitis, 43
classification of rickettsia by, 6—8
Kyasanur Forest disease virus, 42
R. akari, 30
R. australis, 43
R. conorii, 39, 40, 42
R. rickettsii, 41, 42, 87
R. sibirica, 41, 42, 84
Rocky Mountain spotted fever, 40
spirochaetes, 34
typhus fevers, 3
Verruga peruana, 21
Vertebrates
evolution, 55—56
hosts for *R. rickettsii*, 88—91
wild, typhus agglutinating antibody, 121—122
Virales, 12, 13
Virulence, 93, 94
Viruses, see also specific virus, 130
separation from rickettsiae, 12
Visna virus, 130

W

Weil-Felix reaction, 3
Western Montana U agent, 6, 93
Wolbachia, 15, 20, 23
dermacentrophila, 20, 23
melophagi, 3, 20, 23
persica, 20, 33, 34, 76, 81
Wolbachieae, see also specific genus, 7, 13, 15, 20—21, 76
Blattabacterium, 15, 20—21
Rickettsiella, 15, 20—21
Symbiotes, 15, 20—21
Wolbachia, 15, 20
World War I, typhus during, 118
World War II, typhus during, 118

X

Xenopsylla
astia, *R. typhi* and, 116
cheopis, *R. prowazekii* and, 131
R. typhi and, 114, 115, 116, 126, 128—130
Yersinia pestis and, 135

Y

Yellow fever virus, 130
Yersinia pestis, 135
evolution, 2

Z

Zapodidae, 59
Zinssera, 16, 76
R. typhi and, 130

For Product Safety Concerns and Information please contact our EU representative GPSR@taylorandfrancis.com Taylor & Francis Verlag GmbH, Kaufingerstraße 24, 80331 München, Germany